William Everett Cram

Little beasts of field & wood

William Everett Cram

Little beasts of field & wood

ISBN/EAN: 9783743388062

Manufactured in Europe, USA, Canada, Australia, Japa

Cover: Foto ©berggeist007 / pixelio.de

Manufactured and distributed by brebook publishing software (www.brebook.com)

William Everett Cram

Little beasts of field & wood

LITTLE BEASTS
OF FIELD & WOOD

William Everett Cram

Boston
Small, Maynard and Company
1899

University Press

JOHN WILSON AND SON, CAMBRIDGE, U. S. A.

TO

CHARLES CONRAD ABBOTT, M.D.

Preface

ALTHOUGH *practically all the observations referred to in this book were made in New Hampshire, they will, perhaps, on the whole apply equally well to the wild creatures of eastern Massachusetts, and in a more general way to the whole of southern New England and New York. The region which I have chiefly prowled over for the last twenty years — a region, I may add, which two delightful books of New England tales have made, in a way, almost classic — is restricted to the very southeastern corner of the State, and in general character is decidedly not typical of New Hampshire, being simply a quiet rolling farming country, beautiful at all times, but hardly to be called striking or impressive. Its chief claim to beauty is, perhaps, the clearly marked quality of the landscape generally, divided as it is between open fields and pastures, and groves and forests of white pine and* hemlock in solid dark masses, which give it, particularly in winter, a distinctive character not found in regions where the woods are of a more varied make-up.

Ten years ago this region could boast a very respectable area of old growth forest, not the original uncleared primeval forest, to be sure, but the next best thing to it, straight, smooth-stemmed pines whose tops were one hundred feet or more above the earth ; and, between these, shorter hemlocks with dense, almost impenetrable foliage and sturdy trunks three or four feet in diameter which sheltered in their turn the coming generation of saplings springing into life between the prostrate forms of a still older forest.

But by far the larger part of these woods has been cleared since then, and that which remains untouched seems somehow to have lost much of its fascinating wildness. Still as fast as the woods are cleared away new ones are constantly springing up to take their place, so that the actual area covered

PREFACE

with trees has varied less than might be supposed,
while the swamps remain practically unchanged.

The swamps are evidently the basins of what
were once ponds or small lakes. These have become
filled with sediment or covered over with aquatic
growths of one kind and another which in time attained sufficient thickness and stability to support,
first, the bulrushes and alders spreading out from the
shore, and, finally, forests of willows and water-ash
and maple. The tree-roots bound the whole together
and penetrated downward into the water beneath,
now steadily being filled up and confined to slowly
diminishing channels and underground basins, which
I believe are still in existence beneath many of our
swamps to an extent not generally suspected.

I know of one quite extensive swamp in this
vicinity, drained at one end by a brook which for the
first part of its course is roofed over or bridged by
several feet of black loam, which breaks down here
and there for a few yards, and reveals the silent
course of the stream beneath. Although the water

PREFACE

*looks to be only a few inches deep, the sides and bottom
are certainly not well defined, if they exist at all.*

*I had a practical demonstration of this last autumn
when I attempted to step across on a fallen maple
which had evidently been there longer than I had
supposed. It crumbled beneath me, and I went down
until the water reached my waist — but even then I
had failed to touch bottom. By grasping a convenient
sapling, I managed to get myself out without much
difficulty. In doing so, I kicked out beneath the
banks as far as my legs would reach without touching anything more substantial than water-soaked
roots and leaves, and further investigation along the
same brook convinced me that the greater part of
the swamp (which is several miles in extent) is still
in the earlier stages of growth, while most of the
other swamps in the vicinity are much farther
advanced in the process of earth-making.*

*The water courses of this region belong largely
to one type. They have their sources in damp woods,
among low-lying hills and pasture lands hardly fifty*

feet above sea level. But they are lazy, easy-going little brooks that appear always to be seeking the longest route to their final destination, as if loath to exchange the fresh-water meadows and woodlands for the salt marshes, and fearful of losing their identity in the sea. The Squamscott River to the north, and the Merrimack to the south, are merely drawn on a larger scale, hardly differing, except in size, from the little brooks that feed them.

I have watched the daily life of the wild creatures of this locality at all times of year impartially, certainly having spent as much time in the woods in midwinter as at any other season. In describing the habits of the different species, I have endeavoured to give the sum total of what I have seen at different times, instead of trying to depict the life of any single individual of each species.

In grouping my little beasts, I have followed " no law of God or man" beyond associating them as nearly as possible as they are most commonly found associated in their native state. I have begun with

PREFACE

the hunters, the foxes first, and then the weasels, but without including all the weasels in the chapter bearing that title: for the mink and otter I have classed with the swimmers, along with the muskrat; while the skunk, which is also a member of the weasel family, must wait, because of his habits of lethargy, to be classed with the raccoon, woodchuck, chipmunk, and the rest of the hibernators. For a future day, also, I have reserved the wild-mice, — the meadow-mouse, woodmouse, and jumping-mouse, as well as the little foreign pilferers of our store-houses and cupboards whose aspect probably terrorises a greater number of the members of our own species each year than all the bears and wild cats within our borders.

The bats, moles, and shrews must also wait their turn, together with the hares, and, in fact, all of the little warm-blooded furry things which are still to be found within the limits of a day's walk in this part of the country, but for which I have found no space in the present little book.

Even as I now write, I have only to look up from

my desk to see miles of pasture and woodland spread out beneath a blazing August sun that seems to penetrate to the very roots of the grass and the most secret nooks of the forest, as if purposely to reveal anything in hiding there; but I fail to catch even a glimpse of any of the little beasts, though I know well enough that hundreds of them are included within the range of my vision. Only a few minutes ago I saw a woodchuck sitting in his doorway and a chipmunk playing about the roots of a maple, and yesterday, at about this time in the morning, a red fox stood for several moments at the edge of the corn, almost within gunshot of the house, and scrutinised me as inquisitively as I was scrutinising him. In all probability he is hiding at this moment somewhere beneath the shadow of the broad leaves of the corn, waiting, half alseep, for some unwary chicken to wander within his reach.

For permission to reprint portions of certain chapters I am indebted to the kindness of the editors of

PREFACE

the New England Magazine *and the* Popular Science Monthly, — *the greater part of the chapter on red squirrels, as well as portions of the chapters on the mink, otter, and muskrat, having already been published in the* New England Magazine, *while much of the weasel chapters appeared in the pages of the* Popular Science Monthly.

The remainder of the text and most of the illustrations are now printed for the first time.

W. E. C.

Hampton Falls, N. H.
August 15, 1899.

Contents

List of Illustrations

LIST OF ILLUSTRATIONS

Little Beasts of Field and Wood

.

Little Beasts of Field and Wood

Chapter I

Little Beasts and How to Find Them

TO my thinking, the small beasts that still inhabit our woods have been altogether too much neglected by the student of nature, though really much nearer to us and much more easily comprehended than birds, when you have once succeeded in finding them. For that they are more difficult to observe than birds is undeniable.

I am persuaded that most of us would be surprised to learn how many wild animals of the bigness of a cat and upwards pass their lives in the midst of cultivated districts without ever having been seen by men, to die at last of old age, their existence even unsuspected by the owners of the land they dwelt upon.

In studying quadrupeds, the chief thing to bear in mind is that, with the exception of squirrels and woodchucks, and possibly one or two others, all of them have comparatively poor eyesight, at all events for daylight, and apparently not much better for twilight or darkness.

But even with the best of eyes they could only see in one direction at a time, while the slightest screen of grass or foliage conceals everything beyond it. But with a sense of smell and hearing such as theirs, they are instantly aware of anything that takes place in their immediate vicinity, with the exception of the one point towards which the wind blows. And here is where better eyesight would often stand them in good stead, for eyes are more serviceable away from the wind than against it, and the wonder is that in all these generations of hunting and being hunted, their eyes have not reached a higher degree of perfection ; but perhaps any gain in that direction would mean a corresponding loss in the other senses, and so the least important was sacrificed ; and it certainly seems to be true that in no living creature are all of these senses perfect.

While the wind is at your back, you will only get the most unsatisfactory glimpses of any of the fox and weasel tribes; but with it in the opposite direction, you may study them at your leisure ; and to a certain degree this is true of all our wild animals.

In one sense winter is the best time for studying them, for when the snow is in the right condition, you may follow the footsteps of all those that are abroad at that season, and see for yourself just how they have been spending their time. On snow that is twenty-four hours old you can hardly go a dozen rods without crossing the track of one creature or another, and of course they multiply each night so long as the weather is favourable, until in many places it becomes difficult to distinguish between them. Perhaps the best snow on which to study footprints is a good firm crust, not too slippery, with half an inch of fine snow spread on its surface.

Snow that has been blown about a good deal, and then packed by the wind, takes the clearest imprints, showing the exact mould of the feet that made them; but such tracks are apt to be shallow, often little more than scratches, and hard to see at a distance. If the crust is icy and the surface snow very light, most animals slip about on it more or less, often making it difficult to identify their tracks. Very light snow, if more than an inch or two in depth, falls back into the footprint just made, obliterating the outline of the foot and giving the impression of the track of a much larger animal. A damp snow is nearly always satisfactory for tracking, though decidedly unpleasant to walk in; and it often happens that the clearest tracks will be found in snow that has been almost wasted away by the rains.

For some reason or other, the first snow of the season usually shows few tracks upon its surface; perhaps because the feet of the wild creatures have not become toughened against its chill, and they avoid moving about any more than is necessary. At all events, the number of tracks is apt to increase with each successive snow-storm until the last of the season, so that snow in the last of April is sure to present a perfect crisscross of tracks before it is many hours old, partly owing, no doubt, to the hibernating animals, who have nearly all waked up by that time.

When snow is melting rapidly, it is easy to tell at a glance just how long each of the more

7

recent tracks has been made ; but in cold weather it is somewhat more difficult. If the air is not utterly devoid of moisture, you can judge pretty closely by the size of the frost crystals formed at the edge of each footprint. You may also, by taking up a handful of snow around it, tell something from the readiness with which it falls together, but this last method is likely to prove pretty wild guessing, with any but an old hunter or an Indian. In thick woods you must look for hemlock leaves or anything of the kind, and calculate from the comparative frequency with which they occur in the track and on the surrounding snow, and from the strength of the wind and the age of the snow, about how far you are behind your quarry.

But above all things, you must have your eye in readiness to see that which you are not looking for, as on every track there is something for every few rods that can tell you conclusively what you wish to know, if you can only read

8

TRACKS IN THE SNOW

FOX WEASEL MINK OTTER MUSK-RAT RED SQUIRREL GRAY SQUIRREL FLYING SQUIRREL

it aright. It is simply the game played by the detective, and just as intensely fascinating when once you have learned the first few moves. For as the track grows fresher as you follow it, you must stop looking for it at your feet, but away in front of you, for the further you can discern it in its windings among the trees, the more prospect there is of coming upon the one that made it unawares, and with this in view the best track to follow is one that leads you towards the wind.

The snow often reveals curious and interesting things that would otherwise escape notice. Sometimes I have observed that practically all the freshly made tracks in a certain locality pointed the same way, — foxes, weasels, rabbits, squirrels, and partridges, all headed in the same direction without any apparent cause and independent of the season.

The birds of prey in their hunting write the most entertaining histories of their successes and failures on its surface; sometimes just the marks made by the tips of their wing feathers several feet apart on the snow, while half-way between them a mouse track terminates abruptly, though much oftener the hunter plunges deep into the snow in its anxiety to secure its prey.

Last winter I observed where a great horned-owl had dashed at a rabbit and, missing, gone sprawling along the snow-crust, helpless before the velocity of its charge, stripping the leaves from the ground laurel in its endeavours to check its speed, until finally brought to a full stop by the drooping boughs of a hemlock frozen into the snow. Whereupon it regained its feet and walked off a few yards before taking flight, while the rabbit bounded away to cover.

The tracks of a pair of foxes hunting rabbits together make interesting reading, but when weasels have been the hunters, they generally leave such a muddle of intersecting tracks as to make it difficult to follow the course of events.

Many people appear to take it for granted that

the lives of wild animals, especially the rarer kinds and those that are persistently hunted, must be spent in practically unending terror. But judging from what I have seen, I doubt very much if even those that lead the most precarious lives suffer to any great extent from fear. For one thing, I am pretty thoroughly persuaded that even the wisest of them have no actual comprehension or fear of death, and only avoid danger instinctively. Most animals on finding one of their own kind dead in a trap will examine it casually without exhibiting any marked symptoms of alarm, and will perhaps return later to be caught in the same trap, if nothing further arouses their suspicion. But let one of them be caught insecurely and escape, even if but slightly hurt, and that particular trap, or any other like it near by, will be strictly avoided for months, or else carefully approached and sprung in such a way as to make it harmless.

As to the danger that comes to them through

being pursued by other beasts or birds, or by
man, I should say, judging from my own observations, that comparatively few of them experience
more than two or three really narrow escapes in
the course of a year.

Of course this is only guess work, but may in
part serve to correct what I believe to be a generally false impression in the other direction. But
I doubt if it matters very much to the creatures
themselves whether these adventures are few or
many. For danger that brings with it action
and excitement is nearly always found to be, on
the whole, enjoyable instead of the reverse, and
must be doubly so when the knowledge and fear
of death are lacking.

When one animal finds itself pursued by another more powerful, it naturally puts forth all its
energy in order to escape or defend itself; but I
am unable to believe from what I have seen that
in any case it ever enters its mind that the out-
come can possibly prove serious. For it seems

unlikely that any of them, however intelligent, can imagine, except in the vaguest way, anything beyond its own personal experience, and so at most can only dread a few bites or scratches more or less serious. And in times of excitement one never thinks of such things or feels them when they are inflicted, for that matter. And in the case of wild animals, the combat is apt to be short and sharp, and has usually terminated one way or the other before it is time for the pain to set in. That animals frequently escape after having been seized and more or less injured is undeniable, while a still larger number are wounded by shot, but these almost invariably heal with the most astonishing rapidity. I believe that nearly all wild animals forget absolutely whatever fear they may have felt immediately the danger is over, and never think of it again until a similar danger arises, for most of them will resume unconcernedly whatever they were doing as soon as they fancy themselves secure, and there is not a shadow of a doubt in my mind that the sensations of an animal fleeing from a hawk or fox, or even struggling in its grasp, are much the same as those of the foot-ball player or the fencer, neither of whom is exactly to be pitied. So that on the whole it seems safe to conclude that the life of our wild animals is happy in spite of occasional periods of hunger and thirst. Cold appears hardly to affect them at all, except when they are weakened by hunger; and though ill-health is not by any means unknown among them, it would appear to be almost wholly confined to the vege-table-eaters whose food-supply is most abundant and easily procured. The flesh-eaters, probably owing to their more active lives, enjoy apparently unbroken health, though the habit most of them have of gorging themselves to the utmost extent of their capacity after a successful kill would ruin any ordinary digestion; but probably their frequent periods of enforced fasting counteract the effects of over-eating.

Hares, muskrats, gray squirrels, and chipmunks are perhaps the most frequent victims of disease ; while curiously enough woodchucks and skunks, which never take an unnecessary step, apparently, and are enormous eaters at all times, except when dormant in the winter, are astonishingly healthy. Few of the smaller animals seem to be greatly inconvenienced by the loss of a leg after the first few weeks. About the only noticeable change is that an animal with only three legs prefers to escape by hiding wherever it may happen to be, instead of running away, though when necessity requires it, such an one will nearly always be found capable of making as good time as its uninjured fellows.

Although the majority of them are classed as nocturnal, there are very few of our little beasts that do not appreciate the sunlight sufficiently to seek out sunny nooks beneath stumps and ledges for their mid-day naps, even at mid-summer ; just as cats, which are probably about as nocturnal as any of them, are notoriously fond of sunning themselves on every occasion.

And this I believe to be a pretty fair example of the pleasure they find in existence, life for them being evidently divided into but two stages, youth and extreme old age, without the intermediate period of work and responsibility and worry, which we of the present day are taught to look upon as the most important part of human life.

For though there are plenty of hard workers among the rodents, at least, it seems to be work of a wholly irresponsible nature, done for the fun of the thing, as children build forts in the sand. And like children, too, they are certainly fond of pretending. With dead leaves and sticks for play-things some furry little chap will spend hours all alone under the shadows of the leaves, perfectly happy, and, to all outward appearances, just as deeply interested and as much in earnest as when engaged in gathering stores of food for the ensuing winter.

And then they get such an immense amount of satisfaction from the simple matter of eating. Not like birds, most of whom swallow their food whole, and apparently only snatch what fleeting enjoyment they may from its passage down their throats, and the sense of fulness which follows, but like true gastronomers, tasting each morsel appreciatively, even if not much given to lingering over their meals, for the table manners of most of them are, to say the least, decidedly greedy.

Just how much their personal comfort is affected by the weather is of course wholly a matter of conjecture; but I, for my own part at least, have often envied the smaller ones their lot. When tramping across windswept areas of country, without any adequate shelter from the gale for miles perhaps, I have been struck by the abundance of little sun-warmed nooks and hollows where a meadow-mouse, or a hare, or even a fox, might crouch out of the reach of the wind. And

in summer, any bush or broad-leaved herb serves
to protect them from the sun. Almost everywhere, too, there are plenty of holes in the earth,
or the decaying trunks of trees, where they can
find shelter and where neither of the extremes of
heat or cold can ever penetrate.

I should say that by far the greater part of
their discomfort is caused by drought and excessive rainfall or the sudden melting of snow.
There are few sights more pathetic than that of
some little animal whose fur was never meant
to shed water for any length of time, swimming
painfully about among rattling ice-cakes and
sodden snow, with no more cheering prospect
before it in the immediate future than that of
climbing out into the still colder atmosphere, or
squeezing itself away all wet in the damp interior
of a floating log. The sudden overwhelming
of a thunderstorm in summer, when undoubtedly
many of them perish, seems desirable by contrast.

Snow by itself is unquestionably a protection

and benefit to practically all the lesser inhabitants of the woods, and only after it has become saturated with water does it become a menace to their welfare.

It is hardly likely that the droughts we get in this part of the country ever very seriously affect any animal larger than a mole or chipmunk, though a few weeks without rain would probably compel most of the others to seek out new places of abode. But this in itself is not by any means a matter of much consequence to any of them. For I believe that, contrary to general opinion, most of them are decidedly nomadic by preference, and forever wandering from place to place, even when hampered by families of helpless young ones.

RED FOX

Chapter II

Foxes

Red Fox — Black Fox — Gray Fox

IN view of the extent to which the habits of foxes have been chronicled, almost since the dawn of history, it would seem somewhat presumptuous to attempt to add anything new. In fact, the sentence just written strikes me now as hackneyed and worn-out.

But for all that, every encounter with the creature itself, at liberty and living out its own free life as it will, is bound to be a surprise and a revelation to the observer. Not but what the various accounts of fox habits are accurate enough so far as facts go, but all the literature on the subject, it seems to me, has failed signally to render the image of the fox as he really is.

The common red fox of this country is larger and more wolfish than his cousin of the Old

World. His fur is longer and more brightly coloured, and his nose less sharply pointed. Otherwise there seems to be no very striking difference.

My first actual experience with foxes occurred when I was perhaps a dozen years old. I was climbing a thickly-wooded knoll one evening about sunset, when, on coming out into an opening among the young pines, I caught sight of a little yellowish-gray beast as it bolted down into the mouth of a newly made burrow. Supposing it to be a woodchuck or perhaps a rabbit, I let fly an arrow after it, and hurried toward the hole, and then stopped short in my tracks for very fear, because of a cry that issued from the shadow of the trees to one side. I had always been used to hearing the barking of foxes in the distance on still nights in the winter, but this was of so utterly different a character that I failed to associate the two sounds in the least. A harsh, vibrating screech, rising and dying out in a kind of snarling wail, with a weird menacing inflection towards the end, which for the moment scared me beyond any experience I had ever had. And it was getting dark, too, on the east side of the hill, and I had always been a little too fond of reading popular natural history to feel perfectly at ease under the circumstances, so I started for home, but failed to leave that horrible creature behind as I had hoped, for it still skulked along beside me, and its yell rose at regular intervals on the still air, sometimes almost in front.

It was lighter in the more open timber on top of the hill, and here for the first time I obtained a sight of the enemy, following within half a gunshot. I was a good deal relieved to see that it was smaller than its voice had led me to anticipate. So I fired another arrow at it and ran back, causing it to retreat again to the cover of the young growth. But I was not to get rid of it so easily, for when I turned back again, it turned also and followed me to the foot of the hill, still yelling. Most wild animals are bold enough in defence of their young, and the fox is no exception; but I have never since then seen one carry it to quite that degree of recklessness, though recklessness is beyond question one of the prominent traits of fox nature, and its wildness is, I am confident, the result of careful calculation, and not timidity.

It is difficult to understand how so keen an observer as Thoreau should have seen so little of them. I recall but two or three places in all his writings where he mentions having seen one, and in one place he says, " It is remarkable how many tracks of foxes you will see quite near the village, where they have been in the night, and yet a regular walker will not glimpse one oftener than once in eight or ten years." Now it is hardly likely that they are very much more abundant here at present than they were in Concord in Thoreau's day; yet I remember four

that I saw in 1898, in broad daylight, without counting those chased by dogs, and at least as many the year before, and I think perhaps that that is about an average, for most farmers interested in such matters whom I have asked, can recall at least one or more that they have seen each season.

Foxes possess in the highest degree the love of hunting, and under the excitement of the chase are apt to throw caution to the winds. The swallows drew my attention one morning to a fox creeping towards the barn under cover of a stone wall, and evidently intent on mischief. Presently he sprung over the wall at a hen that had strayed away from the yard, and the next instant both came hurtling toward me across the field, the hen cackling hysterically, and the fox bounding along after her in silence, with his yellow fur and bushy tail flashing in the sunlight. It was one of the most beautiful and really thrilling scenes I have ever witnessed, and

I wonder now I had the heart to interrupt it. But I did, and chased the fox away, and then returned to the house for a gun, though without much idea of seeing anything more of the fox, who I supposed was miles away in the woods and still running.

But he was not so badly scared as I had fancied, for on coming back with the gun I encountered him unexpectedly, close to the hen-yard fence, in the act of swinging the hen across his shoulders with its neck in his teeth. It was the regular thing for him to do of course, though this only made it all the more surprising in real life. And he did it so perfectly, too, — just as he is represented as doing in picture-books, and as if he had been in the habit of doing it every day of his life.

I have never seen a more perfect personification of justifiable pride and satisfaction over a successful kill than he exhibited, marching down the garden with the utmost deliberation, as though time was of no consequence to him. And then I felt that it was my turn, for he was hardly forty yards away, and I sighted fairly at his shoulder; but just then my foot slipped on the grass and I sat down very suddenly, while the fox bolted off with his prize still over his shoulder.

By the time I had regained my feet, he was ninety yards away or more; but I sent the heavy shot spinning after him just for the moral effect it might have, and at least it caused him to drop the hen, and put on a fresh spurt of speed for the woods, which in the next instant swallowed him from sight.

From what I have seen, I am inclined to think that the foxes get even more poultry than they have credit for, as they hunt at all hours of the day; and wherever there is a field of corn in which chickens are allowed to wander, you can nearly always find fresh fox tracks in the soft earth, often almost under the farmer's windows.

It is no very uncommon thing to see foxes trotting along at their ease across the meadows, either in summer or winter, and at all hours of the day, and their barking also is often heard in the daytime. I have seen one cross the field on a rainy afternoon in the spring within one hundred yards of my window, barking as he went, and occasionally pausing to sniff — for mice apparently — in the dead grass.

Last June at about ten in the morning of one of the hottest, brightest days we had, I heard the crows cawing uneasily, and looking in their direction I saw a large fox trotting lazily along about a quarter of a mile from the house, though from the direction he was taking he must have been much nearer a few minutes before. As he passed among the cattle, I noticed that each turned its head to look at him, but paid him no further attention.

A little further on, he jumped the brook and stopped for a few seconds to nose about at the edge of the water, or perhaps to drink, and then on again until out of sight. If he kept to the course he was following, he must have gone pretty nearly two miles before reaching the woods, but evidently from choice, for most of the time there were thick evergreen woods within a quarter of a mile or less on either side of him.

On the 20th of last July I heard a general outcry from the robins and thrushes among the sweet-fern and young pines in the pasture, and pushing toward the sound, caught a glimpse of a young fox evidently about to help himself to the contents of a brown thrush's nest just in front of his nose, containing three young birds nearly grown. At my approach he slipped away among the bushes, and I saw no more of him. The young birds crouched down in the nest in silence until I touched them, when one shrieked wildly and scrambled over the edge, and falling caught one foot in the fork of a twig and hung suspended, stopping its outcry immediately. When I replaced it in the nest, it proceeded to clamber over the others and tumble out the other side, while the rest seemed to take it all as a matter of course, and refused to stir.

Two days later as I turned the corner of a clump of young pines near there, I saw an old fox standing perfectly motionless, facing me not twenty yards away in the open, closely-cropped pasture without so much as a thistle to hide behind. It was then about five in the afternoon and the sun shone full on him as he stood on the further bank of a little brook that flowed between us. In that position he had the appearance of being of a very pale yellowish buff, the black on his legs not being at all noticeable, probably owing to the season. I was particularly struck at the time by the size of ears, which were erected to their utmost extent and appeared to come together over the top of his head almost to their tips. After we had eyed each other for a few moments, he swung round and stood side to me, with his eyes still turned in my direction. In that bright sunlight his fur lighted up to a most brilliant red on his back and tail, and was so short and close as to give him a singularly lean appearance.

Presently he turned and cantered slowly away with much of the movement of a trotting horse, slowly veering around toward me however, and coming back along the sheep-path to almost his original position. The birds were scolding him more or less all the time, and a song-sparrow that had alighted in the grass at no great distance seemed to catch his eye, for he at once crouched and attempted to crawl to within springing distance, but without success; then he sat down in the path and proceeded to scratch his ribs with his hind foot, after which he stood up and walked off along the sheep-path toward the woods, every now and then stopping to look back over his shoulder at me as if still in doubt as to my identity. Whenever his head was turned away

I followed, stopping instantly at the first sign of his looking back, and he failed to take alarm so long as he only saw me motionless. When he had reached the shadow of the pines, he trotted up the bank and sat down beside a tree to scruti-nise me; I now intentionally moved my hand while he was looking, and instantly he was off, bounding through the underbrush at a tremen-dous rate.

The next fox that I saw, or at any rate made a note of, was on the 23d of September. It had been showery all the morning, but at noon the sun came out calm and sultry. This fox was pursuing a gray squirrel by the edge of an or-chard; he jumped over a stone wall within fifty yards of me, and stood looking in my direction for several seconds before taking alarm, and then turned and trotted away, while the squirrel ran along the fence and up the branch of an oak-tree. This was hardly one hundred and fifty yards from a farmhouse where men were at work hammering on empty apple barrels; and the fox when he started off ran directly toward them for some ways before turning aside among the trees.

The foxes' den is usually an abandoned woodchuck burrow in a sandy hillside. It is always enlarged and extended considerably so that the big pile of new earth thrown up before the entrance is sure to make it rather conspicuous. The old male, I am told, never enters it, except perhaps to carry food to the cubs; his bedroom is in the open air, usually on a flat rock or ledge, though I have started them from their naps at mid-day on flat stumps in a clearing, among the blueberry bushes in a swamp, or beneath the shelving bank of a stream.

Last winter some hunters captured a fox in the hollow trunk of a large elm that stands alone in the meadows near here. The only opening was through a large hollow root, while inside was a space three or four feet in diameter reach-

ing well up into the tree; and the hunters told
me that to all appearances the fox had not taken
refuge there from the dogs, but had gone in of
his own accord, probably intending to pass the
day there. I once noticed a peculiar-looking
dark-coloured ball five or six inches in diameter
near the entrance of a fox burrow, and at first
was at a loss as to its make-up. But on being
knocked to pieces, it proved to consist of a number of small animals rolled together so com-
pactly as to be indistinguishable at first glance.
I cannot recall exactly the kinds or number of
creatures of which it was composed, but I know
that there were a star-nosed mole, a long-tailed
jumping-mouse, several meadow-mice, and, I
think, a shrew, besides a black-and-white creeping warbler and two or three half-fledged birds
evidently belonging to the same family, some
bits of rabbit fur, and a toad. The young foxes
had evidently been rolling it about in sport, and
it certainly indicated an abundance of food at

the time. I have seen fox burrows where the ground for a considerable space on all sides was fairly carpeted with hen's feathers, and others where partly eaten muskrats and skunks were decidedly in evidence.

Although foxes roam impartially over all sorts of country, high or low, forest or open, they have their established runways, as they are called, where the majority of them travel in going from place to place.

These are not paths like the paths of deer and rabbits, but are in places a quarter of a mile or more in width, and conform to a certain extent to the relative position of farmhouses and bridges; for though fond of hunting by the edge of the water, foxes have a most decided aversion to wet feet, and, except in the winter, have become largely dependent on bridges for crossing even the smaller streams, and their runways seldom pass between houses that are not at least half a mile apart. The course of one of these passes within twenty or thirty rods of my window, and there are comparatively few nights in the year when at least one fox does not pass along it, and often half a dozen, judging from the tracks in the snow. But the fox is a good deal of a tramp, and sometimes for days together there will not be a fox track to be found for miles about.

From what I can learn, fox-hunting, as it is practised in this part of the country, is a science that requires years of study to make it a success. And the novice, though his dogs are of the best, may not even get a shot for the first season or two. Different hunters probably follow different methods; but I know one, at least, who bears the reputation of being the most successful in the region. I don't know how he manages it when the ground is bare, though he seems to be just as successful at such times. But when there is a good tracking snow, he follows the most promising trail, examining it carefully from time to time, until satisfied that he is near his quarry, when he makes a wide circuit, counting all the tracks he crosses in coming round to his former position; and if the one he is following is among them, he makes another circuit, and so on until he feels sure that the fox is inside his last circle. Then he calculates, from the general lay of the land and his knowledge of the ways of foxes, just what course this particular one will be most likely to take when started, and then for the first time allows his dogs to take the track, while he hurries to the position he has fixed upon. Sometimes he finds that while the fox he is after has passed out of his magic circle, another has entered it, or that both may be inside, which increases the interest. If, as is bound to happen at times, the fox starts off by some other than the expected course, the hunter must ascertain which way he is circling, for they seldom follow a direct course, and attempt to head him off at the nearest run-way in that direction. He tells me that in winter a favourite trick of the fox is to lead the hounds on to thin ice over deep water, and that dogs are often drowned in this manner. Although following a totally different style of hunting, and one which is tabooed in England and our own more southern States, he still entertains pretty much the same feeling towards the object of his pursuit, looking upon it not as vermin to be destroyed on all occasions, but as game worthy of being protected at proper seasons. Accordingly, he never allows his dogs to follow a she-fox during the latter part of the winter, and beyond a shadow of a doubt would suffer his last chicken to disappear rather than shoot one in the summer. The fox seems to have little instinctive fear of the dog, and it is not an uncommon thing to see them playing together in the pastures on the most friendly terms; while there are very few dogs that dare to engage an old fox in single combat. Last spring I heard a hound baying on the track of a fox, and soon after saw him emerge from the

woods not far from me, accompanied by a large
St. Bernard, who seemed to be decidedly new to
the business of fox-hunting. The two crossed
over a little bridge together and climbed up the
path beyond among the pines. A little later the
hound's voice stopped abruptly to be superseded
by the shrill barking of a fox, and in another
minute both dogs came hurrying back with the
fox pursuing them. The latter stopped at the
top of the bank overlooking the bridge, while
the dogs crossed over, and then turned to look
back at him. They acted at first exactly as if
puzzled to know just what they ought to do about
it, and then the hound took up the track again
and started towards the fox, baying, while the fox
slipped back out of sight in the shadow.

But in a few minutes the performance was repeated, the hound being driven back to where
his companion stood waiting. And then they
turned together and trotted away in an opposite
direction, leaving the fox victorious. He stood

at the top of the bank watching their retreat, and barking at intervals, until my approach drove him back into the woods. In all probability his family was near there, and he felt it his duty to protect it from invaders.

It must add immensely to the fascination of fox-hunting to know that at any time you may draw that inestimable prize, a black or silver fox. This is now believed to be merely a variety of the common red fox, only with long black fur of an entirely different quality. The white-tipped over-hair varies in amount in different specimens, so that the colour varies from pale silvery gray to absolute black, and other things being equal, the darker the skin, the more valuable it is.

Although occasionally found wherever the red fox is abundant, it is everywhere decidedly rare; and I cannot discover that an undoubted specimen has ever been killed hereabouts. But, a few years ago, all the hunters agreed that

BLACK OR SILVER FOX

W.E.C.

there was one in the vicinity. Most of them
claimed to have seen it; and several to have shot
at it, yet none of them succeeded in bagging it;
and after two seasons of such rumours it ceased
to be talked about, probably having been driven
out of the region.

It seems remarkable that one fox should have
been seen so frequently, yet most of the accounts
appeared trustworthy enough; and I myself, on
two occasions, the first winter saw a fox that
certainly looked quite black. It seems hardly
likely that there should have been more than one
black fox in the neighbourhood at the time;
unless, perhaps, they are in the habit of roaming
about over the country together, attracted by their
similarity of colouring.

Or if, as is said to be the case in most instances,
they were born of red parents, there may have
been several reared in this part of the country at
that time, just as black lambs appear for some
inexplicable cause among the flocks of a certain

4   49

locality during the same season. Last year, for instance, nearly every flock near me had black or brown lambs, though no black or brown sheep had been known here for years. In our own flock were two black twin lambs. One of them had ordinary wool like the rest of the sheep, and from the first showed a slight shade of brown in certain lights, which increased during the summer to a decided brown at the surface; while the fleece of the other was nearly twice as long, and at first more nearly resembled fine silky hair than wool, and for several months was as absolutely black as anything well could be; even now, it only shows a faint coppery tinge at the tips.

Red lambs were also common last year, and in most instances exhibited the same peculiar quality of fleece; but like the brown and some of the black ones, they faded out in course of time to resemble their mothers both in colour and in quality of wool.

Is it possible that the birth of black and silver

foxes is of the same nature as that of the black and brown lambs, and that they have the same tendency to grow paler as they grow older? — for foxes are killed in all the varying shades and qualities of fur, from the black to the common red. One variety known by furriers as the cross fox has the fur of the ordinary colour, while the long over-hair is black, the points falling together on the shoulder in the pattern of a cross.

My theory might be made to account for the fact that the black fox or foxes disappeared so quickly from the neighbourhood, without any having been taken by the hunters.

The first one that I saw crossed the road about one hundred yards from me, and trotted along by the edge of a field on the snow-crust, giving me a good opportunity to observe him, sometimes against a background of dark pines, and at no time did he show any other colour than black, though the sun, which was near setting, was at my back all the time. A few weeks later, on coming

out of some woods into the open, I saw a fox
that looked quite black running before me up
the hill-side; he stopped for a moment to look
back at me, and then disappeared round the hill.
I followed his track in the snow, and found that
he circled round to the woods I had just left, and
from there he refused to be driven, though I followed his track until it crossed and recrossed so
often as to be indecipherable, but I was only able
to get the merest glimpse of him between the
trees. When I saw him first, I was looking
north, and, as it was nearly noon, the sun must
have been at my back, though it may have been
hidden by clouds at the time. Now every
one is aware that an object on the snow looks a
great deal darker than it really is, by contrast;
but I have never seen anything so light as a red
fox look black, unless the sun was directly beyond
it. All of the foxes that I have ever seen when
the ground was bare have been of the ordinary
kind; but other persons appear to have been

more fortunate, and one of these told me that he saw a red and a black fox together.

Every few years it is reported that a black or silver fox has been killed in one or another of the neighbouring towns; but just what proportion of these reports is true I am unable to say.

The gray fox is a wholly distinct species which does not intergrade with the others, being smaller and of decidedly different build. Its colour is dull yellowish gray, and it usually lacks the white tip on the tail so characteristic of the others. Although a southern species whose habitat is commonly given as " Pennsylvania and southward," it appears to possess the wandering habits of its tribe, and, unless I am very much mistaken, is not infrequently found much further north. I have received accounts of two foxes shot within a few miles of me recently, the description of which agreed perfectly with this species. The hunter that killed one of them supposed for a time that he had killed a genuine silver fox.

March 6, 1899, 8 P. M. A fox has been barking, or squalling rather, at no great distance from the house, though in all probability he is farther away than he sounds, that being usually the case. I doubt if at any time he is less than a quarter of a mile away, though his voice sounds about as loud at that distance as the caterwauling of cats in the door-yard, which sound, in fact, it strikingly resembles at times, as if several foxes were fighting and snarling together. At other times it is simply a shrill snarl uttered at intervals of perhaps a minute, with occasional longer intervals; finally he seemed to go off towards the southeast, still barking periodically. *March 7.* It certainly seems as though the fox last night was barking at the approaching storm, as they are said to do, for though it was perfectly clear and calm at the time, before morning it was blowing a gale, with snow-squalls that rendered it impossible to see any distance, and it has been increasing ever since and bids fair to be a blizzard to be remembered.

We have experienced three memorable snow-storms this winter, and before each of them the owls and foxes hooted or barked in a wholly unusual manner, and before the first one, the celebrated November storm, the owls were flying about and hooting before the middle of the afternoon. Now I have never believed the barking of foxes or the hooting of owls to be a reliable forecast of stormy weather on every occasion; but I am fully persuaded that animals are affected by the approach of a storm of unusual violence, and at such times are apt to utter cries unlike their ordinary ones.

Before the storm set in last night, I heard the mice in the walls squeaking as I have not heard them before for months, as if they, too, like the foxes, were excited by its approach.

The footprints of a fox rather resemble those of a small dog, and are ordinarily placed in a straight line, one directly in front of another, and perhaps ten inches apart in the line. Sometimes for short distances, and oftenest in very shallow snow, they are in pairs quite close together, and one slightly in advance, the pairs two or three feet apart. In deep snow they are sometimes in groups, separated by much longer intervals.

The track of a large cat is sometimes mistaken for that of a fox, but the separate footprints of the cat are always shorter and rounder, with the toes gathered together in front and a distinct pad behind, while the footprint of a fox appears to have been made by four pads of equal size, and shows distinctly the marks of the claws in front. And, moreover, the cat's tracks are generally nearer together, and are seldom in a straight line for any great distance.

April 4, 1899. When I awoke this morning I heard the barking of a fox not far away, and he continued to bark at intervals all the time I was dressing. On going to the door I saw him about one hundred and fifty yards away, and at least fifty yards from the edge of the woods.

He was apparently coming towards the house, but the opening of the door must have alarmed him, for he instantly became silent, and, turning, ran a little way towards the woods before stopping to look back at me, and then went loping off across the snow, keeping mostly in the shadow of the woods, though without actually entering them.

The sky was perfectly clear and the sun at least an hour and a half high; but though I was looking towards the east and saw the fox both in sunlight and shadow against a background of snow, I failed to see that he looked much darker by contrast with the white surface, all of which helps to convince me that the one I saw three years ago was actually a black fox.

But to go back to the gray fox. I have never had any opportunity for observing its habits when at liberty. Those who have, speak of it as the in-

ferior of the red fox in every way, except perhaps
in the matter of tree climbing, for which it appears
to be rather better adapted than the other,
though I am unable to learn that it ever does
much more than take refuge among the lower
branches when closely pursued, which is really
no more than the red fox will do on occasion,
though more rarely.

In the open country the gray fox appears to be
at a decided disadvantage. And even in those
parts of the country where it was originally
most abundant, it quickly disappears with the
clearing away of the forest.

That it is a most intelligent animal is beyond
dispute ; but both in intelligence and general
appearance, it seems to me from what I can
gather to be a much more commonplace sort of
little beast, to be classed with the woodchucks
and hares, whose appearance in the woods and
fields seems perfectly natural and hardly likely
to attract especial attention.

GRAY FOX

The sight of the more distinguished-looking
red fox, on the contrary, with his long legs and
brilliant colouring, always fills me with surprise in
spite of myself, as if it were some strange creature
escaped from confinement, and not just a common fox, reared perhaps in my own pasture, and
whose ancestors were dwelling here, for all that
I know to the contrary, ages before my own
crossed the seas. For although certain writers
have tried to prove that ours is only a variation
of the European red fox brought to this country
by the first settlers, there seems to be but slight
argument on their side. And I, for my part, shall
always associate the red fox and the red Indian
as co-inhabitants of the old forests that once
stood here.

ERMINE IN SUMMER

Chapter III

Weasels

WHY is it that while popular fancy has attributed all sorts of uncanny and supernatural qualities to owls and cats, and no ghost story or tale of horrid murder has been considered quite complete without its rat peering from some dark corner, or spider with expanded legs suddenly spinning down from among the rafters, no such grewsome association has ever attached itself to the weasels, creatures whose every habit and characteristic would seem to suggest something of the sort? Now, fond as I am of cats, I should never think of denying that they are uncanny creatures, to say the least. But, suppose it was the custom of our domestic tabbies to vanish abruptly, or even gradually on occasion, like the Cheshire cat after its interview with Alice;

that would at least furnish some excuse for the general prejudice against them, but would really be no more than some of our commonest weasels do whenever it serves their purpose. I remember one summer afternoon I was trout-fishing along a little brook that ran between pine-covered hills. As I lay stretched on the bank at the foot of a great maple I saw a weasel run along in the brush fence some distance away. A few seconds later he was standing on the exposed root of the tree hardly a yard from my eyes. I lay motionless and examined the beautiful creature minutely, till suddenly I found myself staring at the smooth greenish gray root of the maple with no weasel in sight. Judging from my own experience, I should say that this is the usual termination of any chance observations of either weasels or minks.

Occasionally they may be seen to dart into the bushes or behind some log or projecting bank, but much more frequently they vanish with a suddenness that defies the keenest eye-sight.

In all probability this vanishing is accomplished by extreme rapidity of motion, but if this is the case then the creature succeeds in doing something utterly impossible to any other warm-blooded animal of its size. Mice, squirrels, and some of the smaller birds are all of them swift enough at times, but except in the case of the humming-bird none of them, I believe, succeeds in accomplishing the result achieved by the weasels. The humming-bird, in spite of its small size, leaves us a pretty definite impression of the direction it has taken when it darts away; but when a mink, half a yard in length and weighing several pounds, stands motionless before one with his dark coat conspicuous against almost any background, and the next instant is gone without a rustle or the tremor of a blade of grass, it leaves one with an impression of witchcraft difficult to dispel, and best appreciated when one has seen for one's self. Nor is the everyday life of the weasel quiet or commonplace; his one object in life apparently is to kill, first to appease his hunger, then to satisfy his thirst for warm blood, and after that for the mere joy of killing.

The few opportunities I have had for observing these animals have never shown them occupied in any other way, nor can any hint of anything different be gained from the various writers on the subject, while accounts of their attacking and even killing human beings in a kind of blind fury are too numerous and apparently too well authenticated to be entirely ignored. These attacks are said usually to be made by a number of weasels acting in concert, and the motive would appear to be revenge for some injury done to one of their number. There seems to be something peculiar about the entire family of weasels. The American sable or pine marten is said to have strange ways that have puzzled naturalists and hunters for years. In the wilderness no amount of trap-

W·E·C·

ping has any effect on their numbers, nor do they
show any especial fear of man or his works, occasionally even coming into lumber camps at night,
and being especially fond of old logging roads
and woods that have been swept by fire; but at
the slightest hint of approaching civilisation they
disappear, not gradually, but at once and forever,
and the woods know them no more. If there is
anything in the theory of the survival of the fit-
test, why is it that not one marten has discovered
that, like other animals of its size, it could manage to live comfortably enough in the vicinity of
man? The mink and otter still follow the course
of every brook and river and manage to avoid the
keen eyes of the duck hunter, while for six
months in the year their paths are sprinkled with
steel traps set either especially for them or for
the more plebeian muskrat. If a pair of sables
could be persuaded to take up their quarters in
some parts of New England, they could travel for
dozens of miles through dark evergreen woods

with hollow and decaying trees in abundance, and at present there are almost no traps set in a manner that need disturb creatures of their habits. Partridges, rabbits, and squirrels, which form their principal food, are nearly if not quite as abundant as before the country was settled, so that it would certainly not require any very decided change of habits to enable them to exist; but evidently the root of the matter goes deeper than that, and, like some tribes of Indians, it is impossible for them to multiply or flourish except in the primeval forest.

The common weasel or ermine, which is the the only kind I have seen hereabouts, would seem to have everything on its side in the struggle for existence, and when one happens to be killed by some larger inhabitant of the woods it must be due entirely to its own carelessness. Nevertheless, they do occasionally fall victims to owls and foxes, and I once shot a red-tailed hawk that was in the act of devouring one.

Still, these casualties among weasels are probably few and far between. Fortunately, however, they never increase to any great extent. Occasionally in the winter the snow for miles will be covered with their tracks, all made in a single night, and then for weeks not a track is to be seen; but usually they prefer to hunt alone, each having its beat, a mile or more in length, over which it travels back and forth throughout the season, passing any given point at intervals of two or three days. This habit of keeping to the same route instead of wandering at random about the woods is characteristic of the family, the length of the route depending to a certain extent on the size of the animal. The mink is usually about a week in going his rounds, and may cover a dozen miles in that time, while the otter is generally gone a fortnight or three weeks. When it is possible, the ermine prefers to follow the course of old tumble-down stone walls, and lays its course accordingly. In favourable districts he is able to keep to these for miles together, squeezing into the smallest crevices in pursuit of mice or chipmunks. All the weasels travel in a similar manner — that is, by a series of leaps or bounds in such a way that the hind feet strike exactly in the prints made by the fore-paws, so that the tracks left in the snow are peculiar and bear a strong family resemblance. On soft snow the slender body of the ermine leaves its imprint extending from one pair of footprints to the next, and as these are from four to six feet apart, or even more, the impression left in the snow is like the track of some extremely long and slender serpent with pairs of short legs at intervals along its body. I have said that the ermine is the only weasel I have found in this vicinity, but this is not strictly true. One winter I repeatedly noticed the tracks of an exceedingly large weasel — they were so very large, in fact, that I was almost forced to believe that they were those of a mink. The impression of its body in

74

the snow was quite as large as that made by a small mink, but the footprints themselves were smaller, and the creature appeared to avoid the water in a manner quite at variance with the well-known habits of its more amphibious cousin, while, unlike the common weasel, it never followed stone walls or fences. I put my entire mind to the capture of the little beast, and set dozens of traps, but it was well along in the month of March before I succeeded. It proved to be a typical specimen of the Western long-tailed weasel, though I can find no account of any other having been taken east of the Mississippi. Its entire length was about eighteen inches; the tail, which was a little over six, gave the effect at first glance of being tipped with gray instead of black, but a closer inspection showed that the black hairs were confined to the very extremity and were partly concealed by the overlying white ones; the rest of the fur was white, with a slight reddish tinge, and much longer and coarser than

that of an ermine. Since then I have occasionally seen similar tracks, but have not succeeded in capturing a second specimen. In all
probability the least weasel is also to be found
here if one has the patience to search carefully
enough; none, however, has come under my
observation as yet. All the small weasels that I
have seen have proved on close inspection to be
young ermines with thickly furred black-tipped
tails; in the least weasel the tail is thinly covered
with short hair and without any black whatever.
Late in the autumn or early in the winter the
ermine changes from reddish brown to white,
sometimes slightly washed with greenish yellow
or cream colour, and again as brilliantly white as
anything in Nature or art; the end of the tail,
however, remains intensely black, and at first
thought might be supposed to make the animal
conspicuous on the white background of snow,
but in reality has just the opposite effect. Place
an ermine on new-fallen snow in such a way that

LEAST WEASEL

it casts no shadow, and you will find that the black point holds your eye in spite of yourself, and that at a little distance it is quite impossible to follow the outline of the weasel itself. Cover the tail with snow, and you can begin to make out the position of the rest of the animal; but as long as the tip of the tail is in sight you see that and that only. The ptarmigan and northern hare also retain some spot or point of dark colour when they take on their winter dress, and these dark points undoubtedly serve the same purpose as in the case of the ermine.

I base my statement that weasels ordinarily travel by a series of leaps wholly on the appearance of their tracks, being unable to imagine any other mode of progression that could produce them. But at the same time I am bound to confess that I have never seen one moving in that manner. In fact I have only seen three or four living specimens in all, and these only for a few seconds at a time at most, and they seemed to glide along rather, with unarched backs, like serpents, and invariably vanished for good and all before going many feet.

I remember that before I had formed any very definite impression of the nature of a weasel, I was informed by another boy of about my own age, that a weasel had three pairs of legs, which may have been his way of accounting for their peculiar manner of running, for he claimed to have chased and killed two that morning.

Yesterday, March 12, 1899, I found an ermine caught in a box-trap which I had set for the purpose, hoping to keep one in confinement during the spring months in order to observe closely the change from white to red which occurs each spring.

On first looking into the trap I supposed it to be empty; but opening it pretty wide, I discovered a weasel crouched in one corner and partly hidden behind the rabbit's head which had served to entice it into the trap, and I confess that, judging from past experiences, I half expected that it would vanish on the instant and leave me with the empty trap in my hands. But this one made no attempt even to escape by darting out of the opening, though a red squirrel or even a chipmunk would hardly have let such an opportunity slip. So I closed the trap hurriedly and carried it to the house, where I transferred my prisoner to a cage where it could be observed more satisfactorily.

It proved to be a female, and her fur looked as white and thick as that of a mid-winter specimen. What chiefly surprised me was her quietness and apparent docility, there being no exhibition of fear on her part at any time, though her big black eyes occasionally took on an expression of alarm at some sudden and unaccustomed noise or movement of the cage.

Whenever I touched the cage, she would approach as if to examine my hand, but without

6   81

attempting to bite it, seldom making any rapid or impetuous movement, but moving in a lithe, serpentine manner, sometimes with arched back, and at others with her body held nearly straight and close to the floor of the cage.

Almost any other of our small animals under similar circumstances would have flown at the sides of the cage, breaking either its teeth or the wires in its endeavours to force an opening. But the only effort of the kind that I witnessed was after she had been confined for several hours, when I found her trying her teeth on the wires in a most careful and business-like manner, but without much apparent enthusiasm. At my approach, she promptly desisted, appearing to dismiss the project entirely from her mind, though, as subsequent events proved, not wholly giving over her plans for escape by any means.

Being desirous of observing her manner of eating, I took my gun and went into the woods, and was lucky enough to start a gray rabbit within a few minutes' walk of the house, and get back with it while it was still warm. Cutting off its head, I tried to push it through the door of the cage, which, however, proved a little too small. But my weasel showed not the slightest hesitation about coming to my assistance, but, seizing it wherever she could most conveniently, she tugged and shook it until, between us, we managed to get it through.

Then after dragging it to the middle of the cage she returned to lap up the blood spilled near the entrance; after which she drank that which nearly filled one of the ears as it lay with the concave side uppermost; then she turned her attention to the large veins of the neck, appearing to suck them dry before turning away.

By the time she had carefully licked off all the scattered drops from the rabbit's fur, and tasted a little of the spinal-cord where it projected from the vertebrae of the neck, she seemed to have pretty well satisfied her hunger, though she did make one half-hearted attempt to gnaw into the brain at the base of the skull, soon pausing, however, to yawn widely and lick her chops like a cat; and I noticed that throughout the entire meal she had somehow managed to keep herself surprisingly clean and dry, carrying her tail cross-wise over her back or shoulders, as if to keep it off the floor of the cage, which had become uncomfortably wet through the upsetting of her dish of water while we were struggling with the rabbit's head.

She now began to exhibit a desire for sleep, crouching on a dry spot with her chin and throat resting on the floor and her eyes half closed, but without showing signs of curling up, as most small animals do when about to go to sleep.

Remembering the very common belief that weasels and hares sleep with their eyes open, I watched her for some time in silence; and it is true that I did not see her eyes close, except to wink, at any time, and though to all appearances asleep, she always raised her head instantly at my slightest movement. Still, the test was hardly sufficient to prove the correctness of the saying.

I now opened the door into the other part of the cage, which was smaller and filled with dry grass for a bed. After a cautious scrutiny of the entrance, she went in and began vigorously rearranging the grass to suit her taste. So I left her to enjoy her nap in peace, but a few minutes later was startled by a strange and rapid knocking from the direction of the cage, and going to see what was the matter, found her endeavouring to carry the rabbit's head into her sleeping apartment, probably realising that although she was unable to eat any more just then, she was likely to wake up at any time and feel hungry, and determining to have her breakfast within reach, and not be dependent on my tender mercies for the future. But the doorway to her chamber was altogether too narrow; and here she exhibited the first real signs of being

possessed of a temper, apparently being pro-
voked beyond all endurance by her repeated
failures.

Fearing that she might finally succeed, I got a
good stiff wire with the end bent up into a hook,
and getting a firm hold on the rabbit's head,
pulled in an opposite direction, and was surprised
at the strength exhibited by the slim little beast.

Just as often as I succeeded in dragging it
half-way across the cage, she would come whisk-
ing after it, and, twisting loose the wire, go
staggering back in triumph with her prize; and
when I at last got it as far as the doorway,
where I could grasp it with my fingers, she still
struggled with it, sometimes even touching my
fingers with her nose, but never offering to bite
them; though I was rather more careless than
was perhaps wise, for a weasel bite is said sometimes to prove rather serious. Realising at last
that she was struggling against odds too heavy
for her, she concluded to abandon the uneven

contest, and retired to resume her slumbers ; and I saw no more of her for the day.

This morning we found the two cats sitting beside the cage, which they had evidently been dragging about in the night in their endeavours to get at the inmate, who was peering from the door of her chamber, evidently not greatly alarmed by the episode.

When the cats had been driven away, she withdrew her head and spent the rest of the morning apparently curled up in her nest, refusing to come out, though I held tempting bits at the entrance of the cage and squeaked like a mouse to attract her attention.

When I poked her with a spear of grass, she refused to stir or take any notice of me, only once uttering a sharp *chip* like the alarm notes of certain species of warblers. This was the only noise she made, except to hiss softly while struggling with me for the possession of her beloved rabbit-head.

Soon after noon, I found the cage empty, with an opening of about her size forced between the wires at the spot where she had made her first attempt at escape. And as nothing is to be found of her, I am forced to conclude that she made her exit from the room through a mouse-hole in the bottom of a cupboard, and is now in all probability chasing the mice somewhere about the walls of the house.

To-day, I caught a large male ermine in a rat-trap, and, like the last, its fur was as thick and white as in mid-winter.

And for my part, I have never seen one between November and April that showed more than a faint greenish or creamy tint, and this only on the under surface. And I have taken probably a dozen winter specimens at one time and another. And unless my memory is very much at fault, this includes one perfectly white specimen taken in November, before we had

had any snow whatever. An ermine under such circumstances must find its white coat undesirably conspicuous; but I have never so much as caught a glimpse of one at such times, though we often have whole weeks together, even in mid-winter, when the woods are practically free from snow.

Yet the change to white seems to be as complete here, and of almost the same duration, as about Hudson's Bay, for example; though only about two hundred miles to the south of us they are said seldom to turn white at all.

And this peculiarity seems to be just as clearly defined geographically in the Old World, — in Scotland, for instance, the change is said to be complete; while in England, especially in the more southern districts, white specimens are of rare occurrence, though piebald and parti-coloured specimens are not uncommon in winter.

And now to go back to the long-tailed weasel. The different works which I have consulted agree in placing the eastern limit of this species at

Minnesota or thereabouts ; but Thoreau, under date of February 22, 1855, writes :

" T. Farmer showed me an ermine weasel he caught in a trap three or four weeks ago.

" They are not very uncommon about his barns. All white but the tip of the tail. Two conspicuous canine teeth in each jaw.

" In summer they are distinguished from the red weasel, which is a little smaller, by the length of their tails, particularly, six or more inches, while the red ones are not more than two inches long."

Now this description, especially as regards the conspicuous canine teeth in each jaw, and the length of the tail, applies perfectly to the long-tailed weasel and not to the ermine, the length of whose tail is invariably given as three or four inches. Still it seems hardly likely that this family of weasels should have inhabited this part of New England all these years without having been catalogued before ; but one hardly knows what else to think under the circumstances.

It is surprising how rapidly the ermine changes from white to brown after the process is once begun; one that I caught in a trap recently showed the transformation nearly complete, the back being of a peculiar shade of reddish buff with only one or two little spots of pure white fur, while the sides were thickly sprinkled with long white hairs, which were already detached from the skin and constantly shedding. The tail was divided into three distinct sections of colour, black at the tip for about an inch, as in winter, then white for the same distance, and brown next the body. The white of the tail was confined to the long coarse hairs overlying the soft under-fur, which was already brown. The feet were still white, like the under surface of the body and throat, which remain so throughout the season. It was then the 14th of April, and only two or three weeks ago those that I caught showed no sign of changing colour. The new brown fur must have grown out with great rapidity, for it was already about as long as it ever would be.

In fact, I am not yet fully convinced that it was new fur, but rather the old under-fur of the last winter turned brown, and that only the long over-hair is shed in the spring. The more carefully I examined the fur of the specimen before me, the more I was persuaded that this was actually the case, and that the ermine habitually goes about with only its under-fur on during the spring and early summer, and that this is shed late in the summer, to give place to a new coat of short hair, which grows longer and is re-enforced by thick under-fur in the autumn; while the whole turns white in November, through some inexplicable process which works alike with weasel and northern hare and ptarmigan, while the coats of other animals remain practically unchanged as far as colour is concerned.

The nursery where the young weasels are raised is, in most instances, beneath a stump, or

ERMINE IN SPRING (SHOWING CHANGE FROM WHITE TO BROWN)

in the burrow of a chipmunk, probably enlarged and remodelled to some extent; though it is doubtful if weasels of any kind ever dig entirely new burrows,—an undertaking that would hardly seem called for anyway in most sections of country, in view of the general abundance of chipmunks; for not only is the chipmunk's home perfectly adapted to the weasel's purpose, but the rightful owners themselves when caught at home serve to furnish the newcomers with their favourite food.

I recall one instance where an ermine was caught in a trap set in the mouth of a wood-chuck's burrow, and although he may only have been poking about in search of mice or rabbits, which frequently take up their abode in bur-rows which the woodchucks have abandoned, it seems quite as probable that the ermine itself had its home there, for in Europe they are said often to inhabit rabbit warrens, and even the underground tunnels of the mole,— so that one would suppose that a creature so easily suited would find little trouble in obtaining lodgings in any part of this country where burrowing animals of one kind and another are so abundant.

From what little I have seen of them, I should certainly never credit ermines with being possessed of a frolicsome disposition. But they are said, on good authority, to engage frequently in the most grotesque antics and gambollings, apparently for their own amusement, though sometimes even approaching their prey in that manner, as if to allay any suspicion of their evil intention, or possibly in the hope that curiosity may tempt their quarry within reach.

A common habit with them seems to be that of storing up the dead bodies of their victims, after having satisfied their immediate appetites. Dozens of mice and young rabbits and the like sometimes have been found packed away in weasels' dens, something which would account for the fact that while following weasel-tracks about the woods where they have been hunting, one almost never comes across any of the remains of the creatures they have killed, though these are often so much larger than the weasels themselves. For while all weasels undoubtedly live largely upon mice, they seldom appear to exhibit much hesitation about attacking larger game.

It is easy to imagine the ermine starting out on his hunting trips, moving by leisurely, silent bounds over the pine-needles, ready for anything that may turn up. Judging from what I have seen, I should say that he depended largely upon his tireless muscles for success, together with the fact that few creatures are able to find refuge in quarters too narrow for him to follow.

I have known gray rabbits when pursued by ermines to leave the woods and rush frantically out into the open fields, as if aware that their enemy was even better suited than themselves for rapid progress through the thicket and brambles which the rabbit usually looks upon as its chief protection. And it seems as though it knew what it was about in seeking the open at such times, as I have never known the ermine to leave the woods in order to follow it, though really much more a creature of the open fields than the other.

The ermine is particularly fond of white-footed mice, and in winter kills large numbers of them about the stone walls and rotten stumps where the mice have their homes. The chase must be a very exciting one, for the woodmouse is scarcely inferior to the weasel itself in leaping powers, besides being a most skilful and erratic dodger, as any one who has ever tried to corner one will bear witness. The slow, fat-bodied meadow-mice should prove much easier victims; but I have seen but little evidence that the ermine depends on them, to any great extent, for food.

In many places the ermine is said to frequent barns and farm buildings, living on mice and rats and, incidentally, on chickens as well. But those that I have studied have been almost without exception dwellers of woodlands and rocky pastures and brier-grown roadsides. The chief service that they can claim credit for is the destruction of wild mice, and their worst crime the murder of partridges and song-birds.

In summer they catch insects and reptiles, and rob birds' nests by the dozen, and have even been seen to spring into the air and catch birds on the wing, though in all probability most of the birds they capture are surprised on the ground, either by the ermine's creeping silently upon them, or lying hidden in the grass near where they are feeding, until one happens to come within springing distance. At night they must frequently succeed in surprising those that roost near the ground, and pounce upon them before they have time to take alarm.

At times they frequent the banks of small brooks, after the manner of minks, especially those brooks that have rocky shores and a quick current; and it is not by any means impossible that they may occasionally catch small fish in the shallows at the edge of the rapids.

I have never witnessed an exhibition of their swimming powers, but have no doubt that they would prove fairly adept on occasion, being, to all outward appearance at least, much better adapted for it than most of our land animals.

Weasels of all kinds have always borne the reputation of being the most extravagantly cruel and bloodthirsty creatures living; and this is undoubtedly true of them in the sense that, according to their size, they kill more of their fellow-creatures than any other inhabitant of the woods.

But they never appear to exhibit any of that wanton, idle, cold-blooded cruelty inherent in the cat and fox tribes, the various members of which are so fond of deliberately torturing any little beast or bird that is so unfortunate as to fall into their clutches.

The flesh of the ermine, like that of the mink,

consists wholly of bands of dark rubbery muscles laid on rather sparingly, except about the head and neck, where they swell out in a manner that lacks but little of spoiling the general symmetry of form displayed by these creatures. Those that control the movements of the head and jaws, in particular, strike one as almost abnormal in their development, — like those of a bull-dog, for example, — though without displaying any of the clumsiness of structure characteristic of that animal.

MINK IN SUMMER

Chapter IV

Swimmers

Mink — Otter

THE mink closely resembles the otter, but is only about two feet in length. Its habits are similar, but it is much more generally known, the sale of its fur forming no inconsiderable portion of the income of many a trapper and farmer's boy during the winter; for, unlike its larger relative, it has not yet learned to avoid traps to any great extent. Its food consists largely of earth-worms and fish, especially eels, which it captures in warm springs and mudholes where they are bedded for the winter. I know of one spring under the steep river bank where the minks watch patiently until some unfortunate eel is brought into sight by the constant upward movement of the water, when it is quickly seized and dragged out upon the snow. But the struggle does not end here, for when the mink prepares to bear its victim away in triumph, the latter is apt to wind its body around that of its captor, and generally succeeds in throwing him end over end more than once before being finally subdued and hauled away, limp and unresisting, across the snow, which when soft holds faithfully the entire history of the contest, from the first confused and hysterical flounderings at the edge of the water, to the triumphal march of the mink up the steep bank, with the eel dragging alongside.

I have never succeeded in discovering how they go to work to get earth-worms in winter, but that they manage it in one way or another is evident enough, for I have examined the stomachs of half a dozen or more killed when the ground, except around springs and similar places, was frozen hard and covered with snow, and most of them contained large red earth-worms that had been swallowed whole.

If the angler who laboriously chops decaying logs to pieces in order to obtain a few borers for bait for fishing through the ice could learn of the minks how to get such worms as these, he would probably consider himself among the favoured of mortals. But the mink does not always confine himself to such insignificant game, by any manner of means; he not infrequently kills birds and animals as large or larger than himself, neither ducks, partridges, chickens, rabbits, or muskrats being ever wholly safe where minks are abundant.

For the minks are less restricted in their hunting-grounds than the others, especially in winter, when they adopt many of the ways of the ermine and sable, wandering about the woods at random in pursuit of game of any kind, from wild mice to rabbits, travelling with the easy, undulating movement of their family until game is sighted, when they pursue it with rapid bounds and arched back, like a frightened kitten.

Like the otter, they slide down every slope they come to, or worm their way about beneath the snow, like moles, creeping along under fence rails and fallen trees whenever the opportunity offers. They are said to climb trees like squirrels; and one reliable hunter told me that he once shot one in the top of a tall elm; but for my own part I have never seen one do more than clamber about the leaning trunk of a willow, a few feet from the ground.

One winter the minks discovered a swarm of wild bees in the hollow trunk of a fallen hemlock near here, and for a short time simply feasted on bees and honey. When I visited the place, the snow about the log was scattered over with frozen bees and bits of comb; and the one who first told me about it said that he had carried home several pounds of the best honey, and bees enough, as he thought, to stock a hive, but they all died before spring.

A mink will nearly always follow any open

brook it comes to, even if obliged to change its course in order to do so, alternately swimming and wading or walking along the bank. On reaching the limit of the unfrozen water, he will often keep on beneath the ice, especially if the water has fallen away from it so as to leave an air space, and perhaps a narrow strip of turf uncovered along the edge of the water. For it is in just such places that meadow-mice spend the winter, their burrows opening out from the banks in the same manner as muskrat holes. And even the smallest brooks harbour young pickerel and eels, as well as frogs and lizards.

One of the most characteristic traits of the mink is his fondness for squeezing through narrow places, — a feat for which he is especially adapted, for wherever his head can go the rest of his body follows easily enough ; and it is surprising to see how many small passages he manages to wriggle through in the course of a night. Every exposed root or fallen tree, tilted ice cake or stone wall furnishes one or more opportunities which he is sure to improve, to the indignation of the trappers, who declare that if the minks would only be a little more careful of their clothes, their fur would be just as valuable in March as in December. But before the last of the winter most of them have so worn away the glossy over-hair on their shoulders as to have materially reduced the selling price of their skins, besides having recklessly exposed themselves to the increasing sunlight, which causes the tips of the fur to curl over in little hooks invisible to the untrained eye, but instantly seized upon and condemned by the furrier, so that by the time the season for trapping musk-rats has fairly opened, minks are worth only about half as much as they were in the fall, and soon become practically valueless, though with characteristic obstinacy they persist in stepping into traps never intended for them. The trapper who, in November, growled because the musk-

rats kept getting into his mink traps when they still needed several months to become "prime" in, now growls on finding the mink which had so exasperatingly avoided his traps all winter, when his pelt would have sold for several dollars perhaps, caught ignominiously in a muskrat trap in April, his faded coat already showing little tufts of pale under-fur detaching themselves from the skin. For in the spring minks haunt the muskrat grounds more than at any other time, penetrating their burrows and climbing over their cabins, or wading along their paths in the shallow water, and robbing the muskrat traps of their victims.

They never appear to have any established homes, but sleep wherever it is most convenient, often in the nest of a muskrat, after having killed or driven off the owner. A favourite place is beneath a stump, or in the hollow root of a tree, or among rocks, their taste in such matters being very similar to that of the otter.

In April the female raises her family in some such place, or else she digs out a short burrow under the overhanging bank, probably making sure that the entrance, whether natural or artificial, is only just big enough to admit her; for most male animals of the weasel family are addicted to the unfortunate habit of dining on the young of their own race,—a habit which may perhaps explain the pronounced difference in the size of the sexes, the females averaging little more than half as long as their mates, and being much slenderer in proportion to their length; and it seems perfectly reasonable to suppose that when the males first adopted the custom of seeking out and devouring their offspring, the smallest females would have a decided advantage in being able to have their nurseries in quarters so narrow as to make it difficult or impossible for the males to gain admission, so that those families which exhibited the greatest variation in size between sons and daughters would increase the fastest, and the females of the race be smaller with each succeeding generation.

I have no positive evidence that the male minks possess this cannibal instinct, but, about the time the young are brought forth, they suddenly begin to follow the females about everywhere they go, in a way which might be attributed to solicitude for their welfare and a desire to afford them protection, but which I am very much afraid only indicates hunger and a depraved appetite. And the fact that so many of the family of weasels have been convicted of a similar crime is certainly against them.

Trout brooks, tumbling over rocky beds, between wooded hills, are commonly supposed to be the favourite haunts of minks; but they seem to be equally abundant along the more sluggish reed-grown streams and mill-ponds of the low lands, and in tangled swamps, or out on the tide-swept marshes within sound of the breakers, where they usually make their homes in haystacks in order to be above the reach of the tide, digging cosey tunnels in the soft hay, like rat-holes in a haymow in the barn. At low tide they ramble along the bottoms of the intersecting ditches that drain the marsh, where they find an abundant supply of small fry and shell-fish and eels, with an occasional muskrat, bound on the same errand as themselves; or else they slip between the stems of the coarse marsh grass on the borders of the shallow ponds, to waylay the marsh birds feeding there, and pick up the wounded ones left by the sportsmen.

The ditches are a yard or more in depth, and only six or eight inches wide, and are cut at first with perpendicular sides which, after a few years, come together at the top, leaving underground passages where the minks can travel in safety, except in the late fall and winter, when the pro-fessional trapper is on the war-path.

But the trapper has only to place his traps on the bottoms of the ditches, and visit them

between tides whenever it is convenient, know-
ing that twice each day the tide will overflow
them and drown any mink that may have been
so unfortunate as to put his foot into one of
them in the mean time.

Minks exhibit much of the playful humour
of otters, and even when alone are often seen
playing about in the sun like kittens. They
swim rapidly, either under water or on the surface, using all four feet like a dog, and from time
to time raising their long necks and triangular,
snake-like heads several inches above the surface,
to look about them. I have never seen one
chasing fish under water, but have no doubt that
they do it just as otters do, following their prey
through all its frantic twisting and dodging until
it is captured.

Most birds exhibit curiosity rather than alarm
at sight of a mink, or at all events they express
their emotions, whatever they are, in a different
manner. Instead of screaming as they do when

a fox or bird of prey is in sight, they gather in the neighbouring trees without any unusual outcry, while from time to time one of their number flies along slowly just over the mink, as if to examine it, at times approaching quite close, — and then away again to join its fellows.

I once saw a mink trying to catch some robins in an open pasture, and though they were careful to keep out of his way, there was no general alarm given, as would have been the case if the enemy had been a hawk or fox, or even a cat.

In most districts minks vary greatly in abundance from year to year, their numbers depending largely on the value fashion chooses to bestow upon their fur. Some thirty years ago they rose in favour until their skins brought five or ten dollars each, and they were killed in such numbers that in a short time they became almost extinct throughout the country. But, fortunately for them, mink fur soon after became unfashionable again, and remained so for ten or fifteen years, the best skins bringing only fifty or seventy-five cents apiece at that time. And as muskrat fur belonging to the same class was correspondingly cheap, very few traps were set for either animal, so that the minks were able to roam about in comparative safety. At that time I remember that their tracks were to be seen about everywhere in the winter, along the road-sides and by every brook and mill-pond.

For the last few years, however, their fur has been steadily advancing in favour and their numbers have decreased accordingly; and even where they are reasonably abundant in summer and early fall, there are usually few to be found after the first of the winter. Last season, for example, there were several families of them, apparently, in this immediate vicinity, as I repeatedly noticed their footprints about the water, and had two or three good opportunities for observing the animals themselves.

But I heard of at least ten that were killed

within a mile of this place during the last of
October and the first part of November, and, by
the time the snow came, there was not a mink
track to be found for miles about, though until
the streams had frozen to the depth of a foot or
more I found otter tracks after every snow-fall.
The only mink tracks that I have seen this winter
have been in an extensive swamp, two or three miles
to the north, and these only on two occasions.

In December I noticed the tracks of a large
weasel near a brook, and, following them, found
where it had dug down into the snow beneath a
little hemlock and uncovered the dead body of a
female mink. There was no sign, as far as I
could see, that the mink had been shot, or killed
by any of the larger beasts or birds of prey ; it was
simply curled up as if asleep, and I could not
help wondering if it could possibly have been
hibernating, as some hunters claim that they do,
and drowned by the freshet that had flooded the
hollow a few weeks before.

The white spot that so frequently marks the chin or throat of the mink is a peculiar feature, it is so perfectly white and sharply defined against the dark brown fur that surrounds it. It does not seem to depend in any way on age or sex, about one mink in ten being wholly without any white marking whatever. A rather larger number have one half of the chin, as far back as the angle of the mouth, white and the other half brown, the dividing line between the two colours being straight and clear-cut down the exact middle of the chin.

Perhaps three out of four have the chin wholly white, and of these nearly half have a white spot on the throat or breast as well, usually just between the fore-legs, though often in the form of a narrow stripe along the throat, sometimes extending all the way from the chin to the breast or even further, and again broken into a series of spots of varying size, while I have heard of at least one mink marked with a row of spots down its back as well.

These markings, together with the great range of size and colour of adult specimens, for a long time led trappers and naturalists alike to the opinion that there were two distinct species, — a belief firmly held by many, and not absolutely denied by naturalists until comparatively recent times.

They distinguished them as the black or mountain-brook mink, and the common mink or marten, which they claimed was much larger and heavier, with lighter coloured fur and white throat.

Only about fifteen years ago I remember few appeared to doubt the existence of the two species, though how they went to work to classify them as they did is hard to imagine; for judging from scores of skins that I have examined at one time and another, I should say that the smaller specimens are just about as likely to be light coloured with white throats as the largest, and that the white markings are as often found on dark coloured minks as on light ones.

Yesterday, March 10, 1899, I found the tracks of a large mink in some hemlock woods by the edge of a swamp; he appeared to have been chasing rabbits back and forth among the trees, sometimes following in their tracks, and again evidently trying to head them off as they ran in circles; and, judging from the wild leaps they often made, he must have been quite near them at times, though I found no actual evidence of success on the part of the mink.

What surprised me a good deal at first was the fact that he seemed purposely to avoid the water, as there was an open brook only a few rods away which was formerly a great favourite of theirs in winter, since it never freezes and flows for a large part of the way underground, whole reaches of it being roofed over by the trunks of fallen trees and dead leaves and vegetable mould, held together by the roots of the living forest.

But yesterday there were no mink tracks anywhere along its banks, nor have there been any there this winter as far as I know, though I have found them on two or three occasions at no great distance away in the swamp.

It certainly begins to look as if the minks have at last learned that that particular brook is dangerous to them, in winter at least; for there have been traps set there every winter, and all winter, for the last six years or more, and dozens of minks have met their fate there, though I have not heard of any having been taken at that place within the last year or two.

It is a pretty well-known fact among hunters that most flesh-eating animals can be more readily called by imitating the squeaking of mice than in any other way; and it would seem to prove conclusively enough that these creatures depend largely upon the sense of hearing in their struggle for a livelihood.

Standing one day beside an old tumble-down rail-fence that ran along between the woods and

salt marshes, half-hidden among the brambles
and tall grass, I caught the merest glimpse of a
mink slipping along between the bottom rails.
As he was evidently unaware of my presence,
I determined to see more of him, and began
squeaking in imitation of a mouse, and quickly
had the satisfaction of seeing him make his
appearance on a projecting stake much nearer
than when I had first seen him.

Stretching himself along a projecting stake,
he appeared to listen and look in my direction ;
but although I was standing in plain sight on
the edge of the marsh hardly a rod away, the fact
that he was obliged to look directly into the
sun evidently made it quite impossible for him
to distinguish clearly what he saw.

At the end of a few moments he dropped into
the grass and started in my direction, the trembling grass-blades clearly indicating his progress
as he approached nearer and nearer until, almost
at my feet, he vanished, and, in spite of the most

patient waiting on my part, absolutely refused to show himself again.

The tracks of a mink are in pairs, two or three inches apart, and one decidedly in advance, the pairs from one to four or five feet apart. About the only other track with which this one is likely to be confused is that of the musk-rat. But in wet snow the muskrat's toes are seen to be more distinct and separate, and whenever the hind-feet are brought down fairly on the snow, they make a much larger print than the forefeet; while a mink's feet are all practically of the same size.

In deep soft snow, the impression made by the body of the animal serves as a distinction; for the mink's body makes a narrow, rounded mark, while the muskrat's is wide and usually with upright sides and flat bottom. The same distinction holds good whenever the creature burrows into soft snow or forces its way through melting ice at the edge of the water.

But the most distinctive feature is the mark made by the muskrat's tail, which is often dragged continuously, marking a distinct and unbroken line between the footprints several rods in length, and when this is not the case, is nearly always found to touch here and there at intervals, either between the tracks or at one side.

In summer and early autumn when the streams are low and the leaves are at their thickest, minks are particularly fond of prowling about swamps and wet woodlands, keeping to the lowest level with all the pertinacity of running water, and following the drying channels of the smaller brooks for miles together, where the reduced waters just linger along through the depressions in the black mud mottled with whatever of sunlight or moonlight manages to find its way between the leaves.

Here they undoubtedly capture insects and birds in abundance, as well as frogs and small fish. For in times of drought practically all the wild life of the woods congregates about such places, or visits them at certain times of the day, water being one of the things which none of our wild creatures appear willing to go without for many hours at a time.

Along the larger streams the minks keep beneath the over-hanging banks as much as possible, sometimes crawling out along low-growing branches or fallen tree-trunks to watch for fish beneath.

When the alewives run up into the fresh water in June, the minks must find the food problem a very easy matter, as they have only to wait at the edge of the shallows above tide-water until a school comes along, and seize whatever fish are crowded out of the water by their fellows, or become stranded among the pebbles by their own carelessness and impetuosity.

For a few weeks before the alewives run, the suckers are going through practically the same performance, so that for the first few weeks of summer the minks must have more fish at their disposal than they can possibly eat, even though it does come at a time when there are little minks to support.

And I have noticed that the females nearly always choose the immediate vicinity of such places for bringing up their families. These families usually keep together, apparently without wandering to any great extent until cold weather, with the exception of perhaps the old males, who may spend the summer roving about the country at random, as they do in winter and spring, though I frequently find them in company with the others in the autumn, all having been brought together, perhaps, by the abundance of fish or game at some particular spot.

For the last three or four centuries, the race of minks has probably had heavier odds to fight against in the struggle for existence than most of our small animals; and the fact that they are still to be found everywhere, in more or less abun-

dance, even within the limits of many of our
smaller towns and villages, might well be taken to
indicate the possession of superior intelligence on
their part.   But I am inclined to think that they
really owe more to their physical perfection than
to their wits, which, it seems to me, hardly correspond to the agile and vivacious movements of
the creatures themselves, but are rather of the
stubborn, slow-working order, quite incapable of
grasping any new and unforeseen situation, though
never failing their owners in a crisis.

I have seen minks face death in all sorts of
situations, but have never witnessed any exhibition of fear or panic on their part, and am almost
tempted to believe that the race is actually devoid
of any such emotion.

They belong to the tribe of fishers and
hunters, equally at home on the land or in the
water, and ready at any moment to plunge into
the rapids and grapple with fish as large as themselves, or chase and pull down a hare among the

birches, and have inherited their tough and elastic bodies from countless generations of ancestors who gained their living in a similar manner.

While they unquestionably experience the same wild joy in hunting and fishing and fighting as other flesh-eating animals, including man himself, I am unable to discover that they are in the habit of carrying it to the extreme that some of the others do, being content to stop killing when they have satisfied their immediate hunger in most instances.

In warm weather they often leave the fish they have caught lying about on the bank, having satisfied themselves with a few bits from the head or back, or whatever part they find most to their taste. But in winter I have never found anything of the kind, and believe that they are in the way of carrying off and hiding for future use whatever they are unable to eat at the time, though there seems to be no evidence of their storing up the dead bodies of their victims by

the dozen, as their cousins the ermines and pole-
cats are said to do.

When a mink finds a muskrat caught in a
trap, he usually manages to eat pretty nearly half
of it the first night, and returns the following
night for some more. And it is quite possible
that in many instances when the trapper finds his
trap holding the foot of a muskrat only, and
concludes that the latter was determined to get
away, even if it cost a leg, that he has really been
robbed by a mink that has succeeded in dragging
away what was left after his meal.

An old hunter, one of the closest observers of
nature I have ever known, once told me that
female minks hibernated in winter in the same
manner as bears, though it was his belief that,
unlike the bears, they never brought forth their
young at that season. At first I refused to take
the slightest stock in what he said, the whole
thing appeared so absurd and so utterly at variance with the teachings of those naturalists who

have made the closest possible study of the
habits of minks. Since then, however, I have
kept my eyes open for any hint that might have
the slightest bearing on the subject, and to my
surprise have found many things that would
seem to point to the correctness of the old
hunter's theory. To begin with, he said that
late in the winter he had repeatedly known
female minks to make their appearance from
beneath snow that had lain undisturbed for days
or even weeks, the tracks apparently beginning
where he first observed them, the difference in
size between the two sexes being sufficient to
make it easy to distinguish between their tracks
at a glance; and, moreover, since he first began
trapping he had noticed that while the sexes were
about equally abundant in the autumn, the
females always became very scarce at the approach of winter, and remained so until spring,
when they suddenly increased in numbers, and
became much the more abundant of the two.

This is also the experience of trappers in general, and may be verified by any one who cares to take the trouble to look into the matter. Evidently no one has ever discovered a mink in a state of hibernation, — at any rate, no such case appears ever to have been reported; but this does not necessarily prove that it is not a regular habit among them.

The cry of the mink is seldom heard, even in places where they are fairly abundant, as they have evidently learned that the greatest safety lies in silence. It is a peculiarly shrill, rattling, whistle-like scream that can be heard at a considerable distance.

It is remarkable that so large an animal as the otter should still hold its own in comparatively thickly settled districts throughout the country, and be practically unknown to any except naturalists and a few others. I have talked with old trappers who have followed the business for years, and could tell correctly at a

OTTER

glance the age and sex of the creature that made each mink and fox track on the snow, and how long each track had been made; but though they knew that otters were killed in their neighbourhood nearly every season, and that they occasionally destroyed or carried off their traps, they had but the vaguest idea of the animal itself or its habits, nor could they describe or recognise its track in the snow, though it is quite unlike that made by any other creature in these parts, and once seen and recognised cannot be mistaken for anything else. Thoreau in his diary, under date of December 6, 1856, gives probably the best description ever written of otter tracks. He saw them on the ice of Fairhaven Pond and Concord River. After reading it, one seems to have learned all that there is to learn concerning the winter habits of otters. In Thoreau's day, otters were evidently no more abundant or generally known than at present; nor do their habits appear to have changed in any way since then. They still follow in the footsteps of their ancestors, though part of their course may lie between cultivated fields, instead of tangled swamps and forests where trees that had died of sheer old age far outnumbered the living ones. In winter they still coast down hill on the snow crust by moonlight, as they did before the country was settled, but keeping a sharper lookout for steel-traps than formerly, their wariness in this direction at the present day being something wonderful, and probably accounting for the fact of their not having been entirely exterminated. A century or more ago they were very abundant in all parts of the country, but were so persistently trapped and hunted that at last the race seemed on the direct road to extinction. Hunters no longer found their pursuit profitable, and took it for granted that they were extinct in reality, giving them a chance to breathe in comparative safety. At the present day whenever one is killed it has usually strayed by mere chance into some mink or muskrat trap, concealed by freshet or rising tide more successfully than its owner could ever have hoped, or it has fallen to the lucky snapshot of a duck or fox hunter who is hardly aware of the value of his prize when he has secured it.

But the ordinary mink or muskrat trap stands but a slight chance of holding so powerful an animal as an otter; while, judging from a late experience of mine, a shot-gun as ordinarily loaded is not much better.

One week last winter we had a warm rain in the night that carried away most of the snow, and broke up the ice on the streams; and one morning I found the track of an otter within a quarter of a mile of the house, evidently made during the last part of the night.

The animal had gone up-stream, and for the first half mile, which was through a comparatively treeless pasture, only landed two or three times; but farther on, where the stream flowed between

higher banks covered with a dense growth of
young pines, the tracks were much more numerous and wandered farther from the water at times,
showing where the otter had been nosing about
rabbit holes and beneath decaying logs; and here
it was joined by another much more recently
made, which I followed until I felt sure from the
appearance of the trodden snow that the otter
could be only a very few minutes ahead of me.
So I stopped and waited motionless, hoping
to get sight of the animal. In perhaps twenty
minutes an otter came to the surface of the water
hardly thirty yards away, and came swimming
almost directly towards me with the whole outline of his head, back, and tail straight and level
with the water, reminding me of a piece of driftwood pushed along rapidly by the current. After
swimming for a few yards, he sank out of sight,
but almost immediately poked his head up again
through the soft ice still nearer, and for a little
while busied himself wallowing about in the

shallow water among the alders and young pines where the stream had overflowed its banks.

He seemed to progress as swiftly and easily through the water-soaked snow as in the clear water, and sometimes came up beneath soft ice several inches thick, breaking up through it with his whole length at once.

He soon disappeared beneath the ice, to come in sight again out in the open current, swimming down-stream on the further side of the top rails of a fence almost submerged by the freshet, and climbed upon the buttressed trunk of a willow about forty yards from me. As soon as he was fairly out of the water, I fired both barrels at him, and he slid off into the water, but, contrary to my expectations, failed to come to the top again, at least at that place, though a few minutes after I saw an otter swim for a few yards between two willows a hundred yards further down-stream.

I could hardly believe at the time that I had

failed to kill him, for I had number four shot in
my right barrel, and number one in my left, and
a gun that shoots close and hard. But I have
since talked with an old hunter who claims to
have killed several, and he tells me that I might
as well have loaded with sand. According to his
accounts, buckshot is the only suitable size for
shooting otters with, and even then one must get
as close as possible. I certainly hope that what
he says is true, and that the otter is still able to
attend to his fishing.

That was the only time that I am certain of
having seen a live otter at liberty in his native
haunts ; but several years ago I followed one's
tracks until it terminated at a slide on the bank
of a tide-water stream, where a space several
yards wide, reaching from the top of the high
bank to the water, was swept clear of snow by
their sliding.

The top of the bank was covered with trees,
beneath whose roots a kind of natural cavern

had been hollowed out by the water, perhaps a foot high and twenty or thirty feet wide at the entrance, and extending back indefinitely into the gloom.

As I stood there in the margin of the stream examining the tracks, I was startled by a low but very distinct growling not wholly unlike that of a cat with a mouse, though deeper. It issued from the cave above my head, not ten feet away, and continued for several minutes ; but whether the creature was growling at me or at one of its own kind, or over its dinner, I was unable to determine.

The following summer I was fishing with two companions along the fresh water reaches of the same stream, when something entered the water not far from us, with a noise exactly like that made by a good-sized dog plunging after a stick ; but, though only a few rods away, the foliage was so dense that none of us was able to catch so much as a glimpse of the creature that made it, and by the time we had reached the spot there were only the disturbed water and trodden grass to be seen.

The otter subsists almost entirely upon fish, and has the reputation of being a veritable epicure in his preference for certain kinds, catching only salmon or trout where these are to be obtained, and when fish are sufficiently abundant, taking only a few bites from the shoulder of each. But those about here are of necessity forced to content themselves with much plainer fare, though I fancy they get their share of brook trout even now, and probably enjoy them all the more on account of their rarity; and who knows but their ancestors got just as tired of an everlasting diet of salmon and trout as some of our own ancestors are said to have done? Even at the present day the otters probably never suffer from hunger, as there are always plenty of the more commonplace fish to be had, not to mention frogs of half-a-dozen species. And there are indications that point to a still humbler diet of insects; for wherever there are otters, there are sure to be numerous places to be found where they have been clawing up the moss, pine-needles, or turf, as the case may be, often working over several rods in one night. Such spots are often spoken of as playing grounds, and perhaps that is what they really are, as they certainly often look as if made in sport, particularly when, as is often the case, the pine-needles, etc., have been piled in little heaps, with dry and broken sticks showing the marks of teeth scattered about. I have known them to do this sort of thing at all seasons, and even when there were several inches of snow on the ground, but I think that they are much more in the way of it in the spring and fall than at any other time.

The only genuine otter slides that I am familiar with are collected within the space of a mile, though almost every brook has its steep banks here and there down which otters slide from time to time when going back to the water.

But the regular slides, although each may not be used more than a dozen times each year, look after they have been used as if several otters had been sliding there for hours; and there can be little doubt that they do it wholly for the fun of the thing, for I have never found any sign of their having eaten anything at the top of the slide, and those that are most frequently used are not near good fishing-holes, but on shallow reaches of the streams where fish are seldom to be found at any season, owing to the clayey bottom and rapid current. In most instances there is a round-about path from the water to the top of the slide, and sometimes the upper part of the slide itself twists about among the tree-trunks for several yards before the final pitch is reached down the slippery bank to the water.

The incline for the last part of the way is usually something like forty-five degrees, and

the speed attained must be something startling,
though only for a few yards, especially on an icy
snow crust.

The slides that I have been describing, of
which there are four or five, are all on tide-water
streams which remain open long after the fresh
waters are frozen over, which may account for
their popularity. Whenever an otter is travelling across lots through the snow, and comes to
an incline of sufficient steepness, he takes advantage of it and slides to the bottom, just as a fox
or mink will; but I have never known them to
amuse themselves by repeatedly sliding down
the same snow-bank unless there was open water
at its foot.

In cold weather they spend most of their time
beneath the ice, after the manner of muskrats,
and are said to be fond of forcing their way up
little brooks hardly large enough to admit them,
probably driving all the fish before them, or
snapping up those that try to get by.

It is generally denied that otters ever occupy real burrows ; but in nearly every stream there are burrows to be found with the openings under water, like muskrat holes, only a great deal larger, perhaps eight or ten inches in diameter, and not inhabited by muskrats. I have always believed that they were muskrat holes that had been enlarged by otters to be used as occasion offered, when fishing under the ice perhaps, but I have no positive evidence to that effect.

The young are said to be reared in natural caves in the bank above high water, or in hollow logs, or at the bottoms of hollow trees, for in the matter of lodgings they seem usually to prefer putting up with whatever presents itself to exerting themselves in the way of making improvements, — a trait that seems to be characteristic of most carnivorous animals.

MUSQUASH OR MUSKRAT

W.E.C.

Chapter V
Swimmers Concluded

Muskrat

FEW people, probably, realise that if the muskrat were suddenly to become extinct, it would mean a clear loss of something like half a million dollars annually to this country and Canada. For that is the estimated value of the year's crop of raw skins, which has not varied materially in amount since the country was first settled, as the muskrat refuses to be driven off by the most constant persecution, and though at times considerably reduced in numbers in certain districts, a single season suffices to bring them to their former abundance. Wherever there is a stream or pond with reed-grown shores, and yellow-lily roots or clams to be had by diving, muskrats are sure to be found; and their numbers usually depend more on the food-supply than on freedom from trappers. For after a few months of experience with traps, the survivors learn to avoid them so successfully as to render their further pursuit unprofitable, and the trapper must move on to new hunting-grounds — though they generally forget their cautiousness during the succeeding summer, and by autumn are as unsuspecting as ever.

In summer they live in burrows reaching well up into the banks and only a few inches below the turf. When the water is very low at midsummer, they dig canals from the lowest openings to the channel, or perhaps these canals are simply burrows that have caved in, — at any rate, they serve as paths down which the muskrats swim or wade to their feeding-grounds. Most of the burrows are dug in August or September, and at that season I often come upon streams roiled by their digging, and, following up the bank, sometimes for a dozen rods or more, I have discovered them at work at it.

The beginning usually appears to be the most difficult part of the job, as they begin at the bottom of the stream, and, owing to the buoyancy of the water, probably find considerable difficulty in holding themselves down to their work; for they keep rising to the surface, to all appearances completely exhausted, and float about on the water puffing like toy steam-launches, sometimes resting their chins on any support that offers, — a low-growing alder branch, or the root of a tree, or else they climb ashore to dry their fur in the air.

As the work progresses, they come up less frequently, until at last you might watch for hours and only see the loose earth and muddy water pushed out of the hole and swept away by the current. The hole is continued high up into the bank, beyond the reach of ordinary freshets, where it ends in a chamber filled with soft grass and rushes. Several burrows often branch off from this chamber, and as a general thing all their openings are under water, those higher up in the

bank being mostly accidental; but in some instances a door-way is dug out beneath the roots
of a tree, probably for convenience at times of
high water, though used more or less at all
seasons when the stream is free from ice.

Their food varies little during the season, their
main stand-by being aquatic plants and shell-fish;
and these are equally accessible, winter or summer.
The roots of the common yellow water-lily are
several inches in diameter, and yards in length,
looking like great wrinkled green-and-white
reptiles sprawled along the bottom, which is
almost covered with them in places. Inside, they
are white and crisp, but have a watery, character-
less sort of flavour, which appears to suit the
muskrats well enough, however, as they constitute
a very considerable portion of their food.

A muskrat will frequently dig up and tow
ashore a piece larger and heavier than himself, and
settle down for a good square meal; and it is
astonishing to see what an insignificant remnant

is left when he has finished. I know of no other creature of similar size capable of eating so much in a given time, or of spending so much time in eating.

It has always seemed to me that, in his own quiet way, the muskrat enjoys existence as largely as any animal in nature. It is true he lacks the excitement of the chase, which forms so large a part of the lives of the fox, weasel, and cat tribes; but then he almost never suffers from hunger, and any one who has ever watched him enjoying a swim is bound to envy him ever after.

One would suppose that where he is in the way of swimming for every meal he takes, and almost everywhere he goes, he might get a little bored at it and look upon it as work. But it is the commonest thing in the world to see them in hot weather rolling over and over in deep water, and floating lazily about, as if the opportunity only offered once in a life-time.

All summer long they swim and wade and

paddle about in the shrunken streams and ponds, or doze, huddled into a ball, on the edge of the bank, hidden by the rank growth of flags and bulrushes among which they have well-trodden paths leading from place to place. The young are born and brought up in the burrows, but only spend a short time there; for they grow with amazing rapidity, and learn to shift for themselves in a remarkably short time.

Once when I was wading across a stream about knee-deep, a little fellow, somewhat less than half grown, came swimming along close to the bottom; on coming to the place where my boots had stirred up the mud, he seemed bewildered, and rose to the surface to get his bearings, and, after a look around, dived again, and in so doing exposed his tail, which I immediately grasped, lifting him half out of water. He did not appear to be greatly alarmed at the circumstance, and, after some futile attempts at climbing up his own tail to bite my hand, seemed perfectly resigned to await the course of events, resting quietly on the water, which slipped away from his nose as the current swirled past him.

When I released him, he acted as though he had been expecting it all along, and, diving, continued his travels down-stream as if nothing had happened. When the fall rains come, the musk-rats are able to swim about their feeding-grounds, where before they were obliged to walk; and now they begin to have their regular feeding places, to which they carry their roots, or whatever else they find to eat, — anything that they can clamber upon answers, an old log or a tussock, or the ruins of a last year's cabin.

Sometimes they carry a few sticks and pieces of sod to some chosen spot, and build up a resting-place in shallow water. They build their cabins in October and November; and there seem to be about as many methods of building as cabins. Some that I have seen started appeared to be simply solid heaps of sod without any cavity inside at first, but hollowed out later. This, however, would seem a very impracticable way of working when the difficulty of disposing of the material removed is considered ; but perhaps they simply dig down from the top and add whatever is removed to the exterior, for the upper chamber is often close to the surface, with only a shallow covering of material, which allows a passage for the air and sometimes for the inhabitants themselves. The chamber is usually less than a foot in diameter, and lined with soft grass and moss ; a passage extends from this chamber downwards and to one side, to another larger one, more or less filled with water, and from here down to still another, below the bed of the stream itself. The last is formed merely by the junction of several burrows, some of which extend up into the bank, and others to the deepest part of the channel.

This is the commonest type of cabin that I have examined. Some of the smaller ones have

THE CABIN OF THE MUSKRAT

only the upper chamber, without any downward passage whatever; others have it in the centre or near the bottom, with a foot or more of material above; while some that I have seen were large enough to have four or five apartments. One that I saw in a little pond in the woods last November must have been at least four or five feet high, and nearly twice as long. Many of them are built in willow-trees, or on platforms of sticks which the muskrats arrange among the alders; and here they exhibit a good deal of the constructive ability of the beaver, cutting their wood on shore in a similar manner, and often towing it long distances to their building sites, where they wattle it firmly between the alder stems for a foundation.

Cabins so placed are generally composed largely of cat-tail stalks and green twigs, while those on the ground are more often built of mud and pieces of sod. I know of one large one, supported among the alders directly over deep water, that has been in existence at least eight years, and though partly destroyed from time to time by trappers and freshets, has been regularly remodelled each fall, and is still in good condition; while two others near by among the willows have been so interlaced by the fibrous roots of these trees as to be practically indestructible.

They have not changed in outward appearance in the slightest degree for fifteen years or more; and I have not the shadow of a doubt that they are fully twice as old as that, and I am unable to see why they should not last indefinitely. They are perfectly smooth and dome-shaped outside, like the popularly accepted idea of a beaver cabin; and though once or twice the muskrats have attempted to raise a second story, or a lean-to at one side, it has always been washed away sooner or later, leaving the cabin as it was before. One of the cabins appears to be always occupied; but the other, being placed much lower, is under water so much of the time that it has to serve more as a resting-place than a dwelling.

Muskrats will sometimes fix up a hollow stump by merely roofing over the top and having an entrance among the roots beneath ; and I know of one placed on a large leaning willow-tree, where I think the passage-way opens directly into the hollow trunk, which is made to serve as a stairway to the bottom of the stream.

The cabins are not much used except at times of high water and in winter, though I doubt if they are wholly abandoned at any season.

So long as the streams remain well frozen, the muskrat is practically free from care and danger. The temperature about him hardly varies a degree, whatever the weather may be above the ice. He knows nothing of snowstorms or sleet or high wind, while the ice holds firm, though there may be several feet of rushing, foaming water over the ice in times of freshet. Down where he is at work, it flows with the same gentle motion as in summer, barely swaying the water-weed and cresses as it slips between them. He is capable of holding his breath for a surprisingly long time, and when compelled to renew it, has only to come up and breathe it out beneath the ice, where it soon becomes oxygenated by the water. There are always bubbles of varying size just under the ice, but whether they contain oxygen, or only hydrogen gas generated by the decomposing vegetable matter at the bottom, I am unable to say; if the former, they must add materially to the muskrat's air-supply, and, theoretically, contact with freezing water should render them fit to breathe. But there is generally plenty of air to be had close up under the edge of the bank, where the water has receded; and when the muskrat has captured a clam, or succeeded in digging up the root he wants, he swims with it to his cabin, or to his hole in the bank, to devour it at his leisure. Occasionally a foot or more of ice will form

when the water is very high, and when the
stream has finally withdrawn again to its channel
a space is left beneath the ice high enough at
times for a man to stand erect in; and here the
muskrats have room to wander about as they
please, until it settles of its own weight, and even
then the inequalities of the river-bed usually
leave them room enough. These open ways
beneath the ice are of much more frequent occurrence on the little brooks, less than a foot
across, that flow into the larger ones here and
there, and up these the muskrats occasionally
travel, sometimes even to the open springs at
their sources, but not nearly as often as might
be expected; for there is always plenty of food
to be had along these brooks, and room enough
to move about in. But they form the regular
runways of the minks in winter, which would
naturally deter the muskrats to a certain extent;
for at this season minks are almost their only
enemies, and to meet one alone in this narrow

way would be very embarrassing, to say the least. The mink is really the smaller in most instances, but his long snake-like body and wiry muscles give him a decided advantage, and his fondness for muskrat flesh is notorious. This may also in part account for the fact that in winter muskrats rather avoid those shallow portions of the stream that are prevented from freezing by warm springs or a rapid current.

There is such a place near here, where, for nearly a quarter of a mile, the stream only freezes in the most severe weather, and as a consequence has always been a favorite resort of the minks. But though muskrats are often there at other seasons, there is no trace of them to be found there in winter. Still there must be an exception to every rule, and on the sixth of January I have found the tracks of muskrats in the snow at this place. They had been going back and forth along a little muddy brook near where it joins the main stream. One track struck off by itself, and as it was evidently but just made, I followed for several hundred yards, and observed where the animal had climbed the high bank into the woods and dug down into the snow in several places, apparently in search of green stuff, nearly a dozen rods from the water. Then the track turned about and led me back to the stream, appearing again on the opposite bank, and stretching right across the narrow meadow to a little springy bog-hole or quagmire, half-a-dozen rods long and only a few feet wide, and almost free from ice. The muskrat was sitting in the edge of the water here, and we evidently became aware of each other's presence at about the same time. He at once waded out into the water and allowed himself to rest on the dead leaves and sediment at the bottom, with just his back and the top of his head above the surface, and his little black eyes fixed intently on me. The motion of his breathing kept the surface of the water trembling all about him; but he made no other motion of any kind for several minutes; then he began to turn his head slightly from side to side, and at the end of ten minutes evidently decided that I was some kind of inanimate natural object, and sat up in the water to stare at me, then turned about and swam a few feet and dived, but without entirely disappearing beneath the surface, and coming up immediately with a morsel of some kind, which he proceeded to devour.

Diving again, he got hold of something which required his utmost exertion to dislodge; but he finally succeeded, and came up with his face plastered all over with clinging wet leaves and mud, which appeared almost to stifle him, and carrying a big bulrush or flag-root of some kind, with white roots and green inner leaves, firmly grasped in his teeth. He swam with this to the partly submerged root of a tree, where he sat bolt upright and fell to gobbling away at his prize, as if almost famished. After finishing, he leaned over and felt all about in the water for any fragments he might have dropped, and when he found anything raised it to his mouth with one hand. Presently he began to wash himself, after the manner of a squirrel, combing his fur with his claws, and rubbing his face and ears with both paws at the same time.

Any one who has ever handled a muskrat must have observed how loosely they are put together, as if put into a skin several times too large, and hardly attached anywhere ; and this sort of structure certainly has its advantages, for the little chap I was observing deliberately reached his right hind-foot across his back and scratched himself behind his left shoulder. He also appeared able to reach any part of his body with either of his fore-paws, or with his mouth, and would sometimes grasp a fold of skin and pull it around into a more convenient position to work upon. In a surprisingly short time, he looked as dry and fluffy as if he had never been near the water. And then, just as he had completed his toilet, he lost his balance and slipped backward into the muddy water, and had it all to do over again; though it is hard to imagine why he was so particular, for, as soon as he had finished, he went back to his diving and was soon just as untidy as ever.

At first he would pause every little while to stare at me, as if still mistrusting something. But although I had advanced a few steps every time he dived, until I was on the very edge of the ice, and at last crouched so near that I could have touched him with my hand, he failed to take alarm, and finally ceased to pay any attention whatever to my presence. After he had exhausted the supply of roots where he was at work, he swam a few yards and began in another place, where they were more abundant; and when these were gone, he commenced crowding in under the ice at the edge of the water, and backing out again, dragging the roots after him.

He appeared a good deal put out whenever the leaves stuck to his face, and would sit upright and hurl them in all directions with his paws, or slap them off with a side-stroke. His appetite seemed perfectly inexhaustible, and, finally, as my muscles began to rebel at being obliged to maintain the same cramped position for so long without relief, I purposely moved my head slightly, to see what effect it might have on him. He dived instantly with a splash, and sank to the bottom, where the water was about six inches deep, and refused to come to the surface, though I tried to raise him with a stick. There he lay, and as often as I tried to lift him, simply slipped off the stick to one side or the other, in the most aggravating manner, but apparently with no intentional effort on his part. So, fearing that he might persist too long for his own good, I left him in order that he might have a chance to come up into the air and resume his breathing.

The muskrats that dwell in the tide-water

meadows lead a somewhat different life from those I have been describing, at least in the summer. Their holes are back in the high bank at the edge of the woods, and are especially numerous where the river creeps in close to the upland; but where it keeps its distance, they live at the heads of the ditches that drain the meadows, and at ebb-tide follow these down to the main stream, where they hunt for shell-fish until the returning tide drives them back. They also dig temporary holes in the banks at considerable distances from the upland, probably for use at periods of low-running tides. They appear to subsist almost entirely on shell-fish, at least I have never caught them feeding on any of the salt-water plants, and there is not much indication about their burrows of their browsing upon the fresh water and upland vegetation. Where they pass the winter, I am unable to determine. In the section with which I am most familiar, and which extends for a mile or more below fresh water, they nearly always disappear in November or earlier, and seldom put in an appearance again before April. And certainly the salt marshes, with their flood-tides and drifting ice-cakes, would make anything but desirable winter resorts for such ease-loving creatures as muskrats. On the whole, the most probable explanation seems to be that they spend the winter in the fresh waters, though if that were the case the number of musk-rats in the fresh-water streams and ponds would be materially increased, — in fact more than doubled if all those from the marshes stopped as soon as they found suitable conditions for spend-ing the winter. But it has always seemed to me that they were really least abundant in the fresh waters at just the time when they are absent from the salt meadows. In this latitude the ice above the muskrats' haunts commonly remains unbroken for three or four months at least.

On the rivers the ice often breaks up during heavy rains late in the winter, and is swept sea-

ward by the freshet, so that the muskrats see their
protecting roof vanish in a single night, and are
compelled to change their habits abruptly to suit
the changed conditions. But on the smaller
streams and ponds the change is much more
gradual. The ice, by virtue of its smaller
surface, is able to hold its own against the
floods that sweep over it, permitting the muskrats to continue their uneventful life below, until
in March the sun begins to beat down more
directly and for longer hours each day, and
in protected places under the edge of the
evergreens, and especially where the comparatively warm water of spring brooks flows in,
the ice slowly withdraws, revealing a narrow strip
of clear brown water above unfrozen turf or sand,
with water-cress and caddis-worms that somehow
serve to make drifting snow and all its associa-
tions fabulous and incredible.

Rocks and tree-trunks that rise to the surface
collect the heat and melt the ice about them, and

to such places the muskrats resort with their roots and clams, to enjoy them in the open air. Sometimes when the snow is unusually deep in the early spring, and it is difficult even to guess at the whereabouts of well-known streams, the warmth of the earth, prevented from radiating by the thick blanket of snow, raises the temperature of the water so that the ice is melted from beneath. And at such times I have known whole reaches of a stream a rod or more in width to open unexpectedly, showing black smooth water between white banks, accomplishing in a few hours what I had supposed would require weeks at least. As a general thing, the muskrats content themselves with merely climbing to the edge, or occasionally striking off across the snow to the nearest open water or spring hole. But sometimes when there is an inch or less of snow on the ice, they seem to be simultaneously possessed with a desire to run about on its surface; and in a single night the snow will be covered with their tracks, meandering about in a purposeless manner, as if they were only out for the air.

The various openings broaden and extend their boundaries, and run together until the ice is reduced to a rapidly diminishing border along each bank. So long as the streams are kept full by the melting snow and the spring rains, the musk-rats are somewhat restricted in their choice of landing-places, and every projecting fence-rail and stump, or leaning willow-tree, is taken advantage of. As the water recedes, they resort to the tussocks, as fast as these are uncovered ; and when the stream is finally confined again to its original channel, they make their feeding-places beneath the roots of trees close to the water, or on pieces of driftwood still partly afloat, or in hollows in the bank, having wholly resumed their summer habits. They often travel for long distances under water, sticking their noses out from time to time, with a sneeze, for a new supply of air, which is obtained in considerably less than a second, so that at such times they are apt to be mistaken for fish rising for insects. A muskrat presents rather a curious appearance when swimming beneath the surface, the long over-hair being plastered thickly down over silky fur that is still dry and filled with air, which bulges out between the long hairs in the form of glistening bubbles.

I recall some years when the ice melted in long spells of fair weather that would have passed for Indian summer a few months earlier, and when there was not enough rain or melting snow to fill the brooks, which constantly receded between tinkling shelves of ice that dripped continually like eaves in rainy weather; and at such times, of course, the muskrats were compelled to adopt their summer ways almost before the ice vanished from their haunts.

As is said to be the case with beavers, there are here and there muskrats that appear to take a dislike to the social habits of their race, and wander away by themselves, usually following the course of some little brook near the source of which they take up their lonely abode. One of these hermits made his burrow beneath a little bridge in our pasture one spring, and lived there for several months. I frequently saw him swimming up or down the brook, which, except immediately beneath the bridge, was only just big enough for his passage, so that in order to turn round he was under the necessity of climbing out on the bank. When alarmed, he generally sank at once to the bottom, completely damming the stream, which rose and poured over him in a miniature cascade. In swimming down-stream he adjusted his speed to that of the current, and so was able to pass along with comparatively little disturbance; but going in the opposite direction, he made a spectacle of himself sure to attract the attention of any one who might be near by.

The excessive rainfall of the summer of 1898 evidently deluded certain of the more inexperienced muskrats into establishing themselves about merely temporary ponds which are ordinarily only damp hollows, and which must have frozen almost to the bottom with the first severe cold weather. Near the middle of one of these I found a newly-made cabin which appeared to be still inhabited when the shallow water surrounding it became frozen over in November. It was in a little tussocky depression without any permanent water-supply, and half a mile or more from any regular haunt of the muskrats, though forty or fifty rods away there was an isolated frog-pond, surrounded by cat-tails and rushes, where muskrats are occasionally to be found.

The males evidently do a good deal of fighting among themselves in the early spring, each apparently endeavoring to lacerate the tail of his opponent as severely as possible, this member being the one most frequently injured, in most cases showing at least one ugly cut, and oftener three or four, of a pattern only to be made by the chisel-like teeth of a muskrat. They are also occasionally bitten on the back or shoulders, and less frequently about the head, evidently not being at all particular about having all their wounds in front.

I have never had the opportunity of witnessing one of these combats, all the meetings that I have seen between them having been of a friendly nature. Last spring, for example, I was sitting on the side of a hill above a stream that wound around its base, when I saw two muskrats, a dozen rods apart, swimming towards each other, but each evidently wholly unaware of the other's presence on account of the windings of the stream. On meeting, they appeared to touch noses, and then one immediately turned about in the water and swam back down-stream, while the other landed on a tussock of grass with something in its teeth which it proceeded to devour, and which I felt certain it had received from the other.

In swimming, the muskrat depends almost

entirely on its hind-feet and muscular tail for propellers. The tail, which is almost as long as the body and head together, is flattened for the greater part of its length, measures an inch or more in width at the widest part, and is nearly naked. The hind-feet, although not truly webbed, are unusually large and peculiarly adapted for swimming, the toes being so arranged that when the foot is brought forward, they fall back one behind another so as to present but little resistance to the water; but in kicking back the whole foot spreads out, and the spaces between the toes are closed up by close fringes of stiff hair which grow on the sides of each toe and spread out with the back stroke so as to form a most serviceable paddle. The forefeet are small and totally devoid of any swimming apparatus, and I think are usually held tucked up under the throat while swimming. I am not certain just how far the tail is made to serve as propeller, its true office evidently being that of rudder; but it may often be seen wriggling vigorously from side to side, like that of a tadpole, as the animal swims past, and undoubtedly serves to increase the speed.

In swimming on the surface, muskrats often hold the head and shoulders well out of water, and the rest of the body deeply submerged, except for the last half of the tail, which is held up behind several inches out of water and curved over in the form of a hook; in turning, it is brought around strongly to one side like a rudder. They are perfectly capable of entering the water silently and quickly at the same time, and frequently do, though quite as often they go in with a sudden plunge and needless amount of noise which I am inclined to think is intended to warn the others.

Although they are unquestionably fond of meat, I cannot recall having seen one attempt to catch any living animal larger than a clam, except on one occasion, and that attempt proved

unsuccessful. It was on a summer afternoon after
a shower. I was standing on the bank of a stream
when a water-rail flew up from the reeds on the
opposite bank, as if frightened suddenly, and flying with trailing legs across the water, dropped
into the grass on my side of the stream. Before
it was fairly alighted, a muskrat plunged into the
water near the place the rail had started from,
and swimming across, pushed in among the grass
where it had taken refuge, driving it again from
its retreat; whereupon it flew back to its original
position. But the muskrat still followed, and
drove it back and forth several times before giving up the pursuit and swimming off up-stream.

On the other hand, I have seen muskrats let
slip several opportunities for capturing birds of
one kind or another. One frosty morning in
April I was walking along the bank of Old River,
— a quiet, sedgy stream, half swamp and half
mill-pond, winding about and doubling back upon
itself in the shadow of the evergreens, with sev-

eral smaller and equally crooked swampy brooks joining it here and there, each with its own particular fog-bank at sunrise, — when through an opening in the pines I caught sight of a black duck, motionless on the water, staring, with outstretched neck, at a large muskrat swimming past within a yard of him. The ripple caused by the muskrat surrounded the duck and spread beyond him; but the bird's eyes were fixed on the little furry head as it glided along; and when at last the muskrat touched the bottom and waded ashore to crouch on the wet bank and nibble at something it held between its paws, the duck, apparently convinced for the first time that it was only a musquash and not a mink at all, evinced considerable relief and straightway fell to feeding.

At the next bend a muskrat was busily occupied with his breakfast. He first came ashore, towing a sweet flag which he had cut off close to the root. It was very amusing to see him eat it; he began at the bottom, and, sitting up and gripping it firmly with his fore-paws, crowded it into his mouth and swallowed it faster than would seem possible for an animal of his size.

A song sparrow came hopping along the bank towards him, and I wondered if it would try to catch it; but the sparrow evidently did not regard him as a dangerous enemy, for it hopped fearlessly to within a foot of him, whereupon the musquash, having finished the tenderest portion of his flag, entered the water and paddled over to an old stump that stood a few yards from the bank, where he began nibbling at the willow shoots and young hardhack that grew there, but not finding it quite to his taste, came back and, climbing up the steep bank to the foot of a cluster of young pines, fell to browsing on the leaves of wild rose-bushes and sweet-briers. He stood upon his hind-feet, bracing himself with his tail, and with his hand-like fore-paws passed the branch rapidly before his mouth, biting off the clusters of leafy buds as they came along. The birds were now giving warning upon all sides of a hawk in sight, and a large one at that ; but the muskrat, although now at a considerable distance from the water, took no notice of the general outcry, and even when a crow sailed over with high-pitched, angry cawing, refused to look up from his meal, until at last, having apparently satisfied himself, he climbed down to the water and sculled away until hidden by a bend of the river.

Further down, where the stream becomes narrow and deep, and the banks rise abruptly from the water without any fringe of rushes and ·sedge, I saw another swimming along near the opposite bank. On coming to a tussock, he clambered to the top of it, and after snuffing about for a few seconds, carefully lowered himself into the water again without a splash, to repeat the performance at the next tussock he reached ; then, cutting a half-circle out into midstream, he commenced diving in deep water, remaining several minutes beneath the surface each time, and usually coming up about a rod from where he went down. He dived with a peculiarly easy rolling motion, and his fur, which was in splendid condition, looked dry and glossy in the sunlight whenever it was out of water for an instant. He looked at me deliberately several times, but, in spite of the fact that I was walking about openly within fifteen yards of him, showed no alarm at my presence. Whatever he was diving for, he evidently was not very successful, and leaving him still at it I strolled on between the trunks of the pines.

There is a bridge of rough stones across the mouth of Great Swamp Run, but the opening has become partly filled by falling stones and earth, and more water flows over it than beneath it. As I was crossing, I saw a muskrat crouching in the grass on the shore of a little island eight or ten rods away. Presently he stretched himself and, sliding into the water, came swimming in my direction, evidently with the intention of crossing over to the further bank of Old River. Nearer and nearer he came until within ten yards, when he turned about and started off, looking over his shoulder at me as he went; but as I remained perfectly motionless he turned again and came swimming back and forth before me, sometimes hardly a dozen feet away. His fur was in poor condition; there was a spot in the middle of his back where it was almost worn off, and here was a miniature pool of water, a tablespoonful perhaps, held in place by the longer fur and loose skin of his sides, rolling about like quicksilver with his every movement, but never spilling and always full whenever he came up from a dive. He soon began swimming in much smaller circles, as if trying to locate some particular spot, and at last stopped and dived, perhaps hoping to pass beneath the bridge and so out into the river; but presently he came up again on the same side as before. He did this several times, but without success, though one would hardly suppose that he would have much difficulty in finding the opening if there really was any. Before diving he always stopped short and floated for an instant, perhaps to get his lungs well filled with air, and then doubling his head beneath him went down with a plunge. After the third dive he swam almost to my feet and, after looking at me steadily for a few seconds, turned abruptly and swam back to his island, first to a muskrat house that stood there partly hidden in a clump of alders, and then back to the spot where I first saw him. Here he landed and, facing about, stood looking in my direction as if waiting for my departure : so I took the hint and crossed over.

When their homes are flooded by heavy rains, the muskrats may sometimes be seen abroad in great numbers, and at such times they show considerable cleverness in concealing themselves even in the most unpromising situations. One of their favourite tricks on scenting danger is to sink into the water and come up beneath a drifting mass of rubbish, sometimes the merest handful being made to serve. I am convinced that sometimes, when no other concealment offers, one will take a wisp of long grass in its teeth and, stemming the current with a fish-like movement of its tail, allow the loose ends of the grass to drag back over it for protection. They usually do this sort of thing close to the bank where the current is slowest; on being disturbed they dive with a slight splash and reappear a few feet away in the same position, with only the slightest ripple to betray them. At other times, when the wind is blowing, you may see one floating on his stomach with his tail held stiffly up, perhaps an inch above the surface of the water, evidently to serve the purpose of a sail; they manage somehow to hold themselves at right angles to the direction of the wind, and make considerable headway without any visible exertion.

STARTLED

Chapter VI

Squirrels

Red Squirrel

THE red squirrel is eminently practical, for all his crazy antics and nonsense, which to the casual observer might appear to constitute his entire character. A more careful study of his ways will, I believe, convince any one that, unlike the majority of wild creatures, he leaves nothing to chance, though quick to seize on any opportunity that offers to better his condition. The conspicuous white circles about his eyes always give him a rather anxious, startled expression, and when suddenly alarmed he has the most affected way of sitting bolt upright and clutching tragically at his breast with one bony little hand, for all the world like some tragedy queen on the stage. I think he almost invariably presses his left hand to his left side, with claws spread well apart and the white fur puffing out between them. This movement is so habitual with the red squirrel that any one who cares to may see the performance by observing those he may happen to see in the woods or by the road-side. If they would only occasionally clap both hands to their breasts, it would make a much more attractive picture; but this I have never seen them do.

Like the blue jay, the red squirrel is always eager for excitement of any sort, but he has the advantage of not being compelled to wait for circumstances to furnish him with an excuse for getting up a racket. As a general thing, when the blue jays begin shrieking and scolding there is pretty sure to be something at the bottom of it, though it may not be of any great importance; and for that matter I have no positive evidence that the red squirrel ever creates a disturbance without having some object or other to shower his anger upon. But he has a way of choosing such

utterly harmless and inoffensive victims, and of
keeping so safely out of sight without so much as
a chirrup to betray him when any actual danger
threatens, that it is hard to believe that he is ever
more than half in earnest. I have never known
him to take part in the general outcry against a
hawk or owl, although he must encounter the
latter frequently as he rambles about among the
evergreens, their favourite roosting-places, and
would certainly have every excuse for resentment
against them; but in such cases he probably considers his own safety as of the first importance,
and makes a point of retiring as expeditiously as
possible. But let him catch a glimpse of an un-
offending partridge quietly gathering berries or
scratching among the pine-needles, and he immediately pretends to fall into an utterly uncon-
trollable rage. He slowly approaches the bird
with short, scratchy starts, down the tree-trunk,
keeping on the opposite side as much as possible,
and peering out from behind the rough bark and

protecting branches, as if fully aware of his danger and determined on not exposing himself more than necessary, and ever and anon retreating, panic-stricken, back into the shadow, to renew the attack from an opposite direction, barking huskily. Should you approach and flush the bird in his direction, he appears to be thrown into a perfect paroxysm of terror by the whir and rattle of its wings, and vanishes with hysterical chatterings, followed by low murmuring growls from his hiding-place.

It is wholly out of the question to suppose for a moment that he can have any cause for resentment against the grouse family ; and yet, so universal is this habit of scolding and threatening them on every occasion that I find I have gradually fallen into the way, when shooting grouse, of allowing the squirrels to point out my game for me to a certain extent, after the manner of trained pointers, finding that three times out of four I can tell from the way they chatter whether

or not it is a grouse that excites them at the
time. On one occasion, one of them even helped
me to secure a wounded bird as cleverly as a
retriever could have done, although probably
from a different motive. I had made a snap shot
through the hemlocks, and heard the grouse
come to the ground fluttering, but on reaching
the spot I found only a few scattered feathers and
just the faintest possible track on the dry pine-
needles, which I lost completely after following
for a few rods. As I stood there looking for
some clue to guide me, a red squirrel began chat-
tering excitedly a few rods away, hurrying along
from tree to tree, and finally coming to a halt, still
scolding. Thinking it just possible that he had
his eye on my game, I approached, and found
him waltzing madly about among the lower
branches of a pine and glaring fiercely down
into the shadows of a tangled mass of fallen
tree-trunks and branches. I moved cautiously
along beside the windfall, and presently noticed

a place where the spider's web had been parted beside the prostrate bole of a large maple, and reaching in beneath it, drew forth the unfortunate partridge, merely wing tipped, but completely disabled for flying.

The red squirrel has been generally accused of being an inveterate robber of birds' nests, and I am afraid there is a good deal of ground for the accusation; still, I have never observed him in the act of plundering a nest, nor do the small birds generally exhibit any great amount of alarm or anxiety at his presence in the proximity of their homes. In the spring, however, I have seen one persistently chasing pine finches and red polls about the top of a gray birch and putting all his agility into play in his endeavours to catch them, creeping towards them cautiously and cat-like and springing out suddenly when he fancied himself near enough; but the birds always slipped away just in time to save themselves, and although the squirrel persisted in his attempts as long as there were any birds in the tree, I could not help feeling all the time that he really had no idea of succeeding and kept it up only for the fun of the thing, as a kitten hunts dead leaves in the wind. But there is no doubt that these squirrels are extremely fond of raw meat of any kind, with a decided preference for the flesh of birds; and to an animal possessing such tastes, no more tempting repast could be imagined than a nest full of tender fledgelings. Every nest in the woods, high or low, is easily within his reach, and, this being the case, the wonder is that there are enough birds to go round each season. For in the evergreen woods at least there seem to be almost as many squirrels as birds' nests, and every orchard and hard-wood grove is inhabited by them to a certain extent. I once noticed one fast asleep curled up comfortably in a robin's nest, which appeared to fit him as exactly as if made to order and furnished the nicest kind of a cradle. At first I supposed that he must have robbed the nest of its contents and was sleeping off the effects of overeating, but on routing him out and examining the nest I found it to be an abandoned one of the preceding year, and the squirrel innocent, at least of that particular crime.

Occasionally you will see one clinging to the bark of some dead pine or hemlock, and listening, woodpecker-like, to the sounds made by the insects at work beneath the surface. When he has succeeded in locating his prey, he tears off the loose bark with his teeth in great ragged pieces, and presently pounces upon and drags forth a flattened white grub an inch or more in length, which he devours with great apparent relish. He appears to subsist, however, mainly on a vegetable diet, not only fruit, nuts, and berries, but seeds of maples and other trees; and he probably knows of other seeds growing about the woods and swamps, and their various times of ripening. He is a veritable epicure as regards mushrooms, and appears to have some infallible

GATHERING MAPLE SEEDS

rule for distinguishing the edible from the poison-
ous varieties; for he recklessly lunches on those
doubtful kinds usually avoided by the amateur,
the white amanita and some of the pink and
scarlet russulas, for example; and I have never
known him to suffer from such indulgence. But
the principal harvest consists of the seeds of the
different evergreens; and although these vary
greatly in abundance from year to year, there is
generally a sufficient supply of one kind or
another. The white pine is usually rather spar-
ing in its yield; but once in every ten or fifteen
years, perhaps oftener, nearly every tree in the
forest bears enormously, even the younger ones
showing scattered clusters here and there, while
those that have stood for generations present a
roughened, shaggy aspect from the thickly
crowded cones at their summits. At such times
the red squirrels seem determined to gather every
cone before it opens and scatters its seeds to the
winds.

They begin work late in July, while the cones are still green and solid with the milky seeds embedded near the centre and hard to distinguish when the cone is cut open. In the hot July sunshine they hurry about their work, cutting off the cones and tossing them over their shoulders well out beyond the surrounding branches to the ground. Whenever the cone or the twig that supports it is cut or scarred, a drop of glistening, transparent sap oozes forth, turning on exposure to the air to the most tenacious kind of pitch; and it is truly wonderful that the squirrels can manage to keep themselves so clean while engaged in their harvesting. But the majority of them show hardly a trace of pitch anywhere about their persons, though now and then you will run across one with little wisps of fur stuck together, especially about his face and neck and in the longer hairs of his tail, evidently having been particularly unfortunate or careless in his work. Every little while they descend to the ground to bury the cones they have cut off, two or three in a place, covered with pine-needles to the depth of several inches. Probably they have learned by experience just how early it is safe to commence gathering them in order that the seeds may ripen properly; but it is hard to imagine how those that are buried early can possibly escape moulding, especially if it should chance to be a wet season. Perhaps, as in the case of cheese, the flavour is really improved by moulding. At all events, it would seem that the squirrels consider a certain amount of moisture necessary to make the seed palatable, for they never appear to store them in hollow trees as they do nuts and apples, though one would suppose they might save themselves a great deal of extra labour by packing them away on end in some such dry and well ventilated cavity, where the cones would open of their own accord without putting the squirrels to the necessity of gnawing off each scale separately. All through the late summer and fall they keep steadily at work, as long as there are cones to be had by climbing, even after a majority of them have parted with their seeds.

In the winter, even when the snow is several feet deep, the squirrels never appear to have any difficulty about locating their stores, sinking perpendicular shafts down through the drifts in order to reach them. Often, instead of burrowing down repeatedly to each little pile of cones, they dig radiating tunnels along the surface of the ground, from the first one opened to the others near it, dragging the cones laboriously along their winding galleries to the surface and away over the snow to some favourite stump before attempting to open them. To get at the seeds they hold the cone upright in their paws and, beginning at the stem end, bite off the scales at the junction with the core, laying bare two seeds for each scale removed. Long practice has made them experts in the art, and it is surprising to

see how rapidly they manage it. The fragments
of cones cast aside collect about the stump until
it is fairly covered up and buried from sight; and
these mounds of little reddish brown chips are to
be seen scattered about the woods at frequent
intervals, indicating by their presence the comparative abundance of squirrels.

They eat also the seeds of the pitch pine and
spruce; but I am inclined to think that they prefer those of the white pine when these are to be
obtained. The little cones of the hemlock retain
their seeds all winter; so, after the harvesting of
the pine cones is over for the season, the squirrels
turn their attention to these. On still winter days
you may see them springing about among the
elastic branches, clinging to the very tips of the
finely divided sprays at a perilous height in their
endeavours to reach the cones that are hung on
such exasperatingly slender twigs, hardly large
enough even for a squirrel's foot to grasp; and
not infrequently a misstep will send one of them

headlong down towards the earth, usually to save himself by catching hold of one branch or another on the way down. If there should chance to be no branches beneath him, he spreads himself out, like a flying squirrel, as he falls, to a remarkable degree of flatness and strikes so lightly as to escape all injury, even on hard snow crust or ice, and scampers away up the tree again without losing so much as a moment of the time he evidently considers so precious. They usually open the hemlock cones as fast as they gather them, eating the diminutive seeds, hardly larger than a pin's head, at once allowing the scales to fall as they will; and as you stand beneath looking up, these come floating and twinkling down between the branches like snowflakes on a clear day.

The red squirrel's winter home varies according to circumstances. Sometimes it is a complicated burrow beneath a stump, with several apartments and winding galleries; sometimes a hollow branch or woodpecker's hole; while in the ever-

green woods he constructs a nest nearly as in-
genious as the more celebrated one of the beaver.

When convenient, he chooses the nest of some
large bird for a foundation, and in this builds a
structure of moss, bark, pine-needles, and dead
leaves, with walls several inches in thickness, and
a soft nest of dry grass and feathers inside. The
bark used is of two sorts, the rough outer bark
of different trees, broken into small pieces, and
what appears to be the inner bark of the red
cedar, torn into narrow strips or ribbons to bind
the whole together. It is put together with
remarkable solidity, and usually freezes hard
early in the winter, furnishing a thorough defence
against the cold or any other enemy from without. The narrow opening at one side is provided
with a hanging curtain of moss or some similar
substance, easily pushed aside by the inmates,
but immediately falling back into place and
effectually concealing the entrance.

If unable to find a bird's nest situated to their

taste, the squirrels arrange a loose platform or framework of twigs in a convenient crotch, and build their nest on that. There is also a considerable range in the quality of workmanship displayed, in some instances the material being apparently thrown together in the most hap-hazard manner imaginable and even when newly built with an effect of general dilapidation, the work perhaps of young and inexperienced builders.

The young squirrels are occasionally born and reared in these nests, although a hollow tree is usually chosen for a nursery, often merely a low stump, two or three feet high, with the hollow open at the top to all the rains of the season. The interior is filled with a lot of fine dead grass and soft lichens for a bed, which at first thought might be expected to become completely saturated in every shower or rainstorm. But as I now recall the different nests that I have found so situated, I have a strong impression that all of them were sheltered by the overhanging branches of dense hemlock or spruce trees, capable of turning aside the water as effectually as a thatched roof. The young squirrels are most absurd looking little beasts at first, like miniature pug dogs, blind and naked, and with enormous heads. In a few days their fur begins to show like the down on a peach, and as a fringe of short hair along each side of the tail, which at length assumes something of the flattened aspect of that worn by their elders, but without displaying much of the fluffy, shadowy quality of the ideal squirrel tail until late in the following autumn. The fur, from the very first, is so close and dense as to give them the typical red colour of their species, although still so short as to be barely perceptible to the touch, giving them a brilliant, newly painted appearance, like toy squirrels covered with some bright coloured satiny cloth to catch the eyes of children. Although they do not remain long in the nest, they are seldom seen abroad until fully grown, or very nearly so, at least, which is rather remarkable when you come to consider the number that are brought up each summer in every pine grove or thicket where these squirrels are abundant. Occasionally you may see a family of them playing timidly about among the branches, but without display-ing any of the self-confident recklessness of their elders, quick to take alarm at the slightest hint of danger and skurry back into concealment, appar-ently possessing less courage than either the chipmunks or gray squirrels of a similar age.

In March the red squirrels tap the maple-trees for their sap, by gnawing through the bark on the upper sides of horizontal branches. The little cavities so made quickly fill to overflowing, and, stretched at ease, the squirrels regale them-selves to their satisfaction. They also drink the sap that flows from such branches as have been broken or cracked by ice or snow during the pre-ceding winter. But their lives are far too busy to allow them to spend their entire time in this manner, and during their absence the sap is apt

TAPPING THE MAPLE

to form into icicles, which, when the temperature
of the wind and other conditions are favourable,
may be constantly evaporating and gathering new
material at the same time, so that the sugar contained in the sap finally collects in rich, honey-
coloured drops of syrup at the extremity of the
icicle, possessing an even more refined and deli-
cious flavour than that obtained by the more
violent process of boiling. The squirrels appear
perfectly capable of appreciating this fact, and are
pretty certain to be on hand to gather it before it
drops, although often obliged to exert themselves
to their utmost in order to reach it. They have
also learned to take advantage of the downward
flow of sap in the autumn ; but at that season most
of them are so busy with their harvesting that they
can hardly spare much time to it; at all events,
they do not collect it then as in the spring.

In April they turn their attention to the opening blossoms of the elms, and you may see them
hanging to the extremity of the slender twigs,

nibbling away eagerly at what must prove at best
very unsubstantial food. Their position among
the slender and still leafless branches is one that
renders them conspicuous for a considerable distance; yet few persons seeing them so occupied
would recognise them for what they really are,
for from that commanding outlook they are
quickly aware of the approach of any one, and,
instead of attempting to retreat to the larger
branches of the tree for safety, remain motionless
wherever they may happen to be. I have frequently seen several of them scattered about in
the same tree top, without mistrusting at first that
these inanimate looking objects were really alive;
for they have a way of assuming such unnatural and
grotesque positions at these times that I really
believe they intentionally pose as old birds' nests or
the remnant of some last season's caterpillar tent.
After a little while, if they fancy they are not
especially observed, they will usually return to
their repast, swinging themselves from place to

place, at first in a cautious, stealthy manner, with an eye for possible danger.

The red squirrel's diet seems to include pretty nearly everything that is ever eaten by any of our native animals. I have known them to find their way into the pantry of a farmhouse, and sample everything available, appearing to be particularly well pleased with the custards. In the winter, they are sure to be among the first arrivals when the fox or goshawk makes a successful hunt, ready to dispute with the crow and the blue jay for whatever is left after the feast; and in summer they often add grasshoppers and other insects to their *menu*. But in spite of it all, and the fact that they never appear to exhaust the stores of provisions they lay up in the fall, they are invariably lean, without so much as the slightest particle of real fat to be found in any part of their anatomy.

From my own observations I am inclined to give them credit for being far superior to the gray squirrel in intelligence, in spite of their crazy manner and lack of self-control. Commonly the tamest and most familiar animal in the woods, if much hunted they acquire in a very short time a cautiousness only excelled by most creatures after years of constant-persecution. Thus their general abundance is hardly to be wondered at, especially when one considers that they are probably about the healthiest creatures in existence. I have never known an instance of their having been afflicted by any of the diseases common among other rodents.

That the red squirrel is an excellent swimmer is beyond dispute, but for my own part I cannot recall ever having seen him enter the water voluntarily. One autumn afternoon, however, I was sitting high up on the wooded bank of a little stream when one came racing along the opposite shore, close down to the edge of the water, making the dry leaves rustle with a loudness out of all proportion to his size. The dead

leaves extended without a break half-way across
the stream, and in another instant the squirrel was
fairly afloat among them. Without showing
much alarm at his predicament, however, he
turned, and swimming in a half circle, as easily
as a mink, with just the top of his head and
a narrow strip along his back and tail in sight,
landed a few yards further down, when for the
first time his unfitness for that sort of thing became apparent. A mink or muskrat on emerg-
ing from the water gives himself a shake and in
a very few minutes is as dry and furry as ever.
But this particular squirrel was literally wet to
the skin, and the more he shook himself the
worse he looked, for while here and there a tuft
of fur attempted to resume its former position,
by far the larger part of it remained flattened close
to his hide, the effect of which his pitiful little
shred of a tail only heightened as he scrambled
over the water-soaked roots and up the trunk of
an ash-tree, where, seated on a projecting knot,

he endeavoured to put himself into better shape.

Occasionally, in the late fall or early winter, there are days when the red squirrels are to be seen everywhere running about over the ground instead of among the branches, both in the woods and in the open. I am inclined to think that such days are usually followed by rain, though whether or not the movement has anything to do with the desire for migration which from time to time seizes upon most of our wild animals, is hard to determine.

Like most northern animals, the red squirrel is rather lighter coloured in winter than in summer, though hardly enough to render him less conspicuous on the snow. As cold weather comes on he changes from a strong reddish brown slightly grizzled with black to a soft grayish fawn colour possessing a most decided shade of green on the sides and flanks, and with a broad stripe of intensely fine light red down the back. The black stripe dividing the dark fur above from the white below, which is often so conspicuous in summer, is usually entirely absent in the winter, when, perhaps owing to the greater length of the fur, the line of demarcation is much less distinct and regular. The colour of the tail is practically the same at all seasons. I have often wondered what use the long hairs springing from the wrists of these squirrels could possibly be to their owners. These slender, whisker-like hairs, often an inch and a half or more in length, surround the wrist in a kind of whorl, and may, perhaps, give the squirrel timely notice when his foot comes within grasping distance of an object, as he leaps madly about among the branches with his eyes on other things.

The red squirrel is apparently affected less by changes of temperature than the other members of his tribe, for he may often be seen hard at work in the hottest weather, and again out on the snow crust at sunrise in the extreme depth of a cold wave. And when, as so often happens, the latter is followed by a sudden thaw, with thick white fog-banks lurking under the evergreens, while the snow sinks beneath the slopping rain and everything is nasty and uncomfortable, the indomitable red squirrels are pretty certain to be out everywhere, not only in the woods but along every roadside fence and stone wall, especially where certain neglected apple trees still hold out an inducement in the shape of frozen apples softened again for the first time in months.

The red squirrel is decidedly a northern species, hardly to be considered native south of the latitude of New York, but ranging towards the pole as far as the woods extend. Unlike the European squirrel, however, which it so closely resembles in many ways, its fur, even in the most northern latitudes, though thick, soft, and of very presentable length, never appears to attain that peculiar and indescribable quality demanded by furriers, a circumstance for which the red squirrel should be duly thankful.

Chapter VII
More Squirrels

Gray Squirrel — Flying Squirrel

THE gray squirrel is apparently much more sensitive to weather conditions than his red cousin, and regulates his movements more in accordance with the season, coming out in summer early in the morning and spending the greater part of the day in concealment, to appear again for a few hours late in the afternoon. As the weather grows cooler in the autumn, he rises later in the morning and curtails his midday nap, so that after the first of November, or thereabouts, the best time to find him abroad is between ten o'clock and noon.

Of course this applies only in a general way. Wild animals are not to be bound down to fixed rules, but come and go pretty much as they please; and individual tastes differ among them as among men.

In my immediate vicinity the hard woods are restricted to certain comparatively limited groves and thickets, and the gray squirrels are of course naturally confined to these, although here and there is a family reared in the evergreen woods, subsisting probably on berries and mushrooms and seeds of one kind and another. But in the autumn they move to the hard woods, to the hickories by preference, and when these fail to yield a crop, to the white oaks or chestnuts or beeches, according to the year; for none of these trees can be depended upon to bear each season, and the gray squirrel population drifts about from one locality to another, assembling in considerable numbers wherever food is most abundant, collecting all the nuts that are to be had and storing them beneath stumps and in hollow trees by the bushel. These stores they live upon until they are finally exhausted, when they move on again; unless the same trees should continue bearing for successive years, in which case the squirrels are likely to settle down indefinitely or until a short crop starts them off again. As a wet summer is thought to blight the blossoms of the hickory, but to have an opposite effect on the chestnut and, I think, the acorn crop, one may to a certain extent judge from the character of the summer where to look for gray squirrels in the following autumn. Where the various kinds of nut-bearing trees are associated about equally in the same forest, I am inclined to think that certain families of squirrels establish themselves and occupy the same hollow trees for generations, the only movement being the occasional influx from less favourable districts, and the departure of the younger ones when the colony is threatened with overcrowding.

The vast migrations which formerly gave this species the name of migratory squirrel, seem now to be restricted to the unimportant wanderings above referred to, but when the country was newly settled they were of not infrequent occurrence, and compared favourably in magnitude with the well-known swarming of lemmings across northern Europe.

The red squirrel is popularly supposed to drive away the gray variety, and probably does to a certain extent, for he is pretty sure to attack the other on sight and generally comes out ahead, although an actual hand-to-hand tussle is of rare occurrence, the encounter generally consisting of ill-natured bickerings at a distance of ten inches or more, terminating in the retreat of the larger of the two combatants. I have been told, however, that when the gray squirrel is fairly cornered, he usually succeeds in putting the other to flight, and at all events I have never known the red squirrels to succeed in actually clearing any grove, no matter how small, of the enemy, although often outnumbering them three to one.

The gray squirrel's home, as already mentioned,

is commonly a hollow tree or branch, some particularly capacious interiors harbouring a dozen or more individuals at certain seasons, and, strangely enough, without any very noticeable quarrels, although the old males are apt to be unpleasantly ugly and tyrannical. They also construct arboreal nests like those of the red squirrel, only smaller as compared with the size of the builder, composed of broad leaves cut off while still green in the late summer, half a dozen in a bunch adhering to the twig they grew upon. These are placed in successive layers on a slight platform of twigs in such a manner as to shed water satisfactorily enough, but without leaving much space inside, even for a single inmate.

I watched the construction of one of these nests for nearly the whole of a hot afternoon. The squirrel, a big dark-coloured fellow with a splendid brush, kept hurrying out to the ends of the branches to clip off twigs, a quarter of an inch in diameter, apparently with a single stroke of his incisors, and then back to his nest with the fan-like cluster of leaves waving above his head. In spite of the squirrel's general distaste for hot weather, this one worked uninterruptedly for hours in a temperature of something like eighty or eighty-five degrees in the shade, with not another of his kind to be seen or heard in the vicinity. Judging from what I saw at the time, I concluded that the leaves were merely arranged in a solid mass, and that the chamber was formed afterwards by the squirrel forcing his way into the centre from one side, and without any attempt at a lining whatever. In this vicinity they are usually placed in beeches and made of leaves from the same tree, whether the beech-trees happen to be bearing that year or not. Although of such frequent occurrence, I cannot learn that they are used with any degree of regularity. I remember that when gunning with other boys it was the custom, when nothing better offered, to fire at these nests whenever they looked promising, but they invariably proved to be uninhabited at the time, although we tried it at all seasons and at all hours of the day. On one occasion, however, I found a nest that contained inmates. It was a little before noon, of a cold brilliant windy day about the first of January. I had jammed a shell of number four shot in my gun and was unable to extract it. Wishing to substitute a smaller size, I fired at the nearest squirrel's nest, and was surprised at the rumpus that followed my shot. After a few seconds a gray squirrel backed out of the entrance, ran along the branch for a few yards, and dropped dead into the snow. I climbed to the nest and found another squirrel inside, which, I think, escaped uninjured.

The barking of the gray squirrel is a decidedly striking sound, audible in calm weather for an eighth of a mile or more, and usually expressive of anger, alarm, and warning. It consists first of a succession of flat, rasping quacks, finally drawn out and prolonged to a kind of whining snarl, rising at times so as to approach the screaming of a hawk in quality. It is heard oftenest directly after rain, when several of them may often be heard answering each other from different parts of the forest.

I have seen them at different seasons running about at the edge of the salt water, at a considerable distance from the woods, apparently for the sake of tasting the salt. I am convinced that most of our wild animals have the same habit when the salt water is accessible, for I have seen a woodchuck leave the woods, go down to the edge of a salt pool left by the tide, and for several seconds lap eagerly at the whitish scum formed by the evaporation of the water.

Gray squirrels are frequently tamed, and are said to make most intelligent and entertaining pets, although rather too much inclined to insist upon having their own way about things. When not confined in a cage their tricks are pretty certain to prevent the lives of their owners from becoming flat and monotonous. One of the most amusing cases that has come to my knowledge recently, is that of a squirrel which persisted in hiding the nuts that were given him in a little girl's hair, and continued bringing them and tucking them away out of sight as long as she would sit still for him to do so.

The gray squirrel is not much seen during the summer months, but in August one will begin to see the young ones, rather more than half grown, going about in pairs, or even three and four together, especially late in the afternoon as the sun gets low, or when the air has been cooled by a shower. They are seldom accompanied by their parents, and have evidently learned to shift for themselves, gathering whatever of fruit or berries the forest has to offer and eating it on the spot, without attempting to put by any surplus for future use. But when the summer is over, and the nuts begin to ripen, they follow the ex-

ample of their elders, and after the first severe
frost of autumn they make the stiff leaves of the
nut trees fairly crash as they leap from branch to
branch, in the hurry of their harvesting. Down
below, the ferns droop and blacken in the open
places in the woods, the scent of frost-killed
vegetation hangs like incense on the still air, and
the bees seek out the banks of goldenrod and
asters for the last honey of the season.

Of all the inhabitants of the forest, the squirrels,
both red and gray, appear to be the least suscep-
tible to the doom of autumn, the vague, unrea-
soning sadness and sense of looking backward
which pervades everything and gives the same
meaning to the notes of migrating birds, the
cricket's creaking, the sound of the wind in the
pines, or the surf beneath the sand-dunes. But
although to all outward appearances the fall is
the squirrel's favourite season, it is also their time
of greatest danger, for the red-tailed and the red-
shouldered hawks are on their migrations, and the

hours that they habitually choose to spend in hunting correspond exactly with the squirrel's working hours, from seven to ten o'clock in the morning and from three o'clock in the afternoon until near sunset. They watch cat-like for an opportunity to take some unhappy squirrel unawares, or, circling above the tree tops, their keen eyes penetrate the foliage from constantly varying positions, searching branch and bole and the carpet of fallen leaves beneath, till, perceiving the flicker of a bushy tail, the long wings close, of a sudden, fan-like, and the hunter goes down with a rush to match his quickness against the quickness of a squirrel. Or the still more treacherous goshawk and cooper's hawk, with their shorter wings and slender yacht-like build, shoot along with baffling swiftness through the undergrowth just in order to surprise the busy harvesters at their work.

When the nuts are all gathered or fallen, the gray squirrel spends most of his time indoors, coming out in the warmest weather to enjoy the sunshine and rake over the dead leaves for scattered nuts and acorns, or to transfer some of his hidden treasures to the home-tree. And in the winter he oftens allows whole weeks to slip by without so much as having poked his head out of his doorway, his favourite time for taking the air being on still sunny mornings after a snowstorm, the least breath of wind often serving to drive him back to his quarters again. Occasionally, however, you will see him out defying the cold when the north wind is fashioning snow-drifts along the fences. But for all that he is no true northerner, and one is hardly surprised to learn that New Hampshire is about his northern limit.

These northern gray squirrels of ours, true to the general rule, are larger and grayer than those farther south. The brown stripe along the back, a distinguishing feature of the typical gray squirrel, is inconspicuous or wholly wanting. The black

variety, formerly the most abundant in the West
and South and even now not uncommon in many
districts, as far as I can learn is practically unknown in New England. I have occasionally
seen specimens that looked quite black in the
distance, but the darkest of those that I have had
the opportunity of examining close at hand
proved to be merely very dark gray above and
the colour of rusty iron beneath, with an unusual
amount of grizzly black about the tail.

The black squirrels appear to have steadily
diminished in numbers since the country was
first settled, and no wonder, for they must find
concealment difficult at all times and doubly so
in the winter, — though possibly in regions
where forest fires are of frequent occurrence the
blackened wood may serve to render them somewhat less conspicuous. But under almost any
circumstances, one would suppose that the gray
squirrel would have the advantage when it came
to a question of hiding from enemies, and it is

rather surprising that the black ones managed to hold their own so well against the Indians and hawks for untold generations. In the primeval forest, however, hiding must have been a comparatively easy matter, and perhaps the blackness of their fur served them in good stead in ways we know not of. For, if we can believe the scientists, the mere instance of a majority of squirrels thinking black fur more beautiful than gray, and so being attracted by it in the mating season, would alone be enough to offset a long list of dangers.

I have sometimes wondered just what law of ownership exists among squirrels regarding their hidden stores: if they really possess any sense of honour in the matter, or whether, as on the whole seems more probable, each has to depend on his skill at hiding and defending his treasures against all comers.

And when a squirrel is killed, how long a time, I wonder, is likely to elapse before his stores are discovered and appropriated by other squirrels or by mice?

The number of the gray squirrels is reduced each year much more rapidly than is the case with the red ones. I doubt if more than one half of them live to see the first snowstorm. And the possessions which they leave, in the shape of nuts and acorns, must help materially in furnishing the survivors with food at just the time of the year when it is most needed, although in all likelihood by far the larger part of them falls into the hands of the red squirrels, whose ranks are never noticeably thinned and who are wide-awake and abroad at all times and seasons.

It might very naturally be supposed that the habits of so curious and remarkable an animal as the flying squirrel would have received more than usual attention, yet, in spite of its abundance and familiarity, there would seem to be less known concerning its ways and manner of getting a living than concerning those of almost any of our wild animals, and for my own part I am almost beginning to despair of ever finding out anything more on the subject.

From what little I have seen, I should class them as creatures of singularly erratic habits, sometimes dwelling alone in hollow stumps close to the ground, or else high up in some deserted crow's nest, and again congregating in communities of twenty or thirty in the hollow trunk of a decaying sapling only a few inches in diameter and scarcely large enough to accommodate them. Farther south we frequently hear of their taking up their quarters in the walls of a farmhouse, after the manner of mice, but no instance of the sort has ever come under my immediate notice, nor can I now recall ever having read of anything of the kind taking place in this part of the country. Here they appear to keep themselves to the thickest parts of the woods, and I have found them most abundant in places removed at least a mile from any dwelling. My father has told me

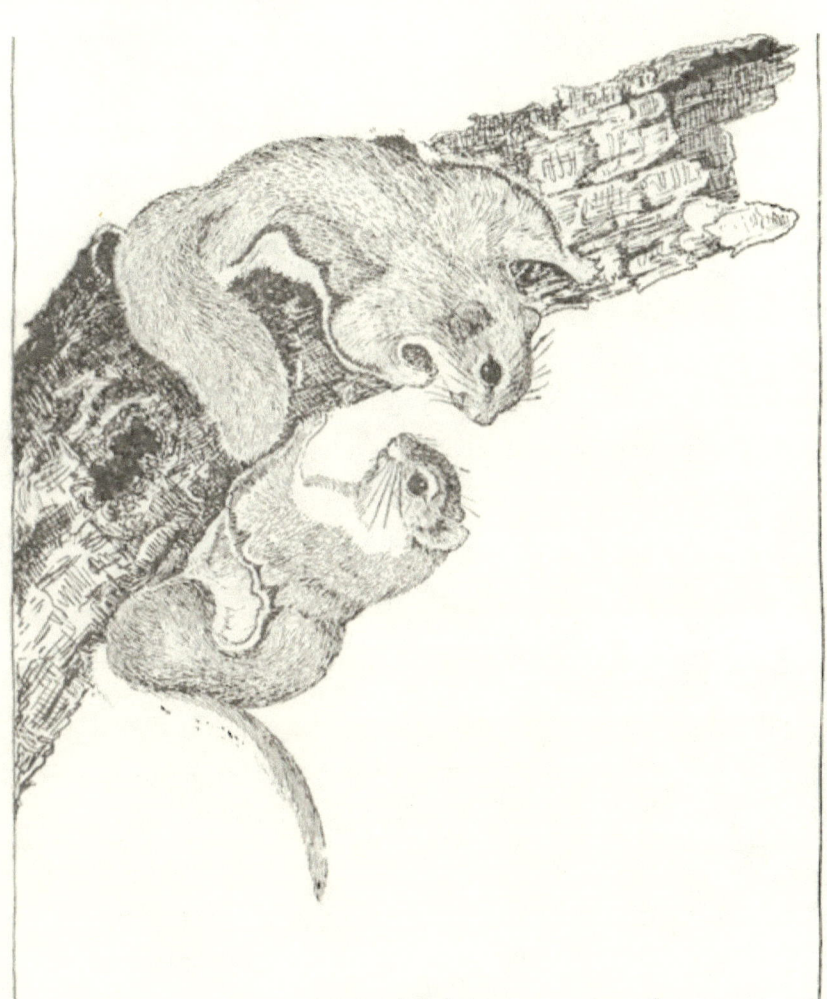

that thirty or forty years ago a colony of them
inhabited a hollow linden, one of a scattered
group of half a dozen trees standing in the open
pasture several hundred yards from any woods.
The old tree is still standing, but, as far as I
know, it has not harboured a flying squirrel for at
least a dozen years.

The last opportunity I had for observing flying
squirrels occurred five or six years since. As I
was tramping through some high rocky woods, I
noticed what looked like a newly made woodpecker's hole near the top of a small dead ash,
and attempted to climb to it. The tree proved
to be pretty thoroughly rotten, and swayed about
a good deal with my movements. By the time I
was half-way to the top, a little round head was
poked out of a hole above me, and soon after a
flying squirrel emerged and scrambled round to
the other side. He was followed by another, and
this one by still another, until half a dozen or
more had made their appearance and clung

motionless about the loose bark, none of them apparently looking at me, but straight before him in whatever direction he happened to be facing, as if his thoughts were of other things. One that was almost within reach of my hand seemed to be quite blind, for the centre of each eye showed a perfectly opaque white spot like the eye of a blind horse. But he appeared as well able to take care of himself as were any of the others, although if I remember rightly he refused to leave the tree and finally crept back into the hole from which he had emerged. The others sailed off one by one to other trees or to the ground, along which they ran like chipmunks.

All the flying squirrels I have ever seen under like conditions have behaved in this manner, apparently acting wholly upon instinct and without displaying the slightest symptom of intelligence; but for all that, there are no more attractive or winning creatures in the woods. They never exhibit any marked symptoms of fear, but just cuddle up on a knot or projecting piece of bark only a few feet away, looking as if they would like nothing better than to be taken in the hand and petted.

I remember hearing my grandmother tell how one winter evening she was sitting before the fire, when my grandfather came home from the woods and taking off his coat threw it across a chair near the fireplace. Presently a flying squirrel crawled out of one of the pockets, sailed across the room to where she sat, and nestled contentedly in her hair, which she wore in a great fluffy mass piled high above her head. I cannot recall the sequel of the story, which was undoubtedly interesting, at all events to those chiefly concerned in it. No one ever knew exactly how the squirrel came to be in the coat, but it was supposed that a family of them must have been disturbed by the choppers in the wood-lot and that this one had taken refuge in my grandfather's pocket, probably bereft of what little wit it ever had by the noise of chopping and the crash of falling trees, and glad to find any retreat away from so rude a world. Perhaps it was only half awakened from its winter's sleep, and dozed off again as soon as it found itself finally ensconced in the depths of the pocket, to be aroused later by the heat of the fire. I cannot help wondering what finally became of it, and just how much of an impression the adventure made upon its sleepy little brain, or whether it took it all as a matter of course, to be forgotten as soon as it was fairly back in the trees again. Perhaps I have run across some of its descendants in the woods or caught them in box-traps without mistrusting that their ancestor and mine had once been on such very intimate terms.

I have never at any time seen a flying squirrel abroad in the day time of its own free-will, even in the darkest weather, and should not hesitate to class it as wholly nocturnal in its habits, although, when routed out in the daylight and compelled to fly, it seems able to distinguish pretty clearly objects several rods away. Its eyes are unusually large and prominent and perfectly black, and its fur is of much the same quality as chinchilla, and of even softer tints. The flying membrane consists of a thin strip of skin stretched between the fore and hind legs and furred above and below. A slight cartilaginous support runs back from the wrist, assisting to extend the membrane when the fore legs are spread apart as in flying. The tail probably serves both as parachute and rudder, since it is thin and flat but of such a close, silky texture as to catch the wind like a sail.

I am not even sure just what the track of flying squirrels looks like, though I frequently run across a track which I suppose to belong to them, as I know of no other animal that could very well make it. But if it really is a flying squirrel track, then they are in the habit of being out on the snow in mid-winter much more fre-

quently than is generally supposed, for I as
often find them freshly made after a long spell of
zero weather as during a thaw. This track may
be best described as in size and appearance intermediate between that of the red squirrel and
white-footed mouse, occasionally showing the
footprints spread well apart laterally, as might
be expected of a flying squirrel. The creature
that makes them, whatever it is, appears to
ramble about the woods and swamps pretty much
at random, climbing low bushes here and there
for seeds or frozen berries of one kind and
another, which are generally eaten as soon as they
are gathered. But if these really are flying
squirrel tracks, then in one way they indicate a
certain degree of intelligence which I had hardly
supposed them to possess, for they never lead
me directly to the home tree, usually terminating
at the foot of some tree or sapling quite devoid
of any cavity or nest.

I should imagine that from their nocturnal

FLYING SQUIRREL

habits these squirrels would fall frequent victims to the different kinds of owls, although I cannot recall ever having found any evidence of this having been the case, either about the nests or in the stomachs of such owls as I have examined. But in all probability they are frequently snapped up by them, as well as by foxes, weasels, and the like, just as they are occasionally by domestic cats.

Oddly enough, one rather frequent cause of their destruction is the barbed-wire fence, the sharp points of which catch the loose skin of their parachute as they sail along. In their struggles to free themselves, the unfortunate squirrels simply twist themselves up tighter and tighter, and in all probability die from suffocation. I have found four or five of these unhappy victims suspended in this way, and I have no doubt from their positions that that they perished in the manner described. I have also seen a fox, a skunk, and several cats caught on wire fences in the same way. The fox was undoubtedly caught while attempting to jump through, but the others may possibly have been shot and hung there by their destroyers.

The colour of the flying squirrel is a pale blue-gray, more or less washed at the surface with buff or fawn colour, or clouded with dusky. From what little I have been able to gather, I should say that the fawn colour is most in evidence in southern specimens, which also appear to be the smallest, while the northern ones, those taken here in New Hampshire, for example, are larger and darker than the type, as if approaching the Canadian form, which in turn approaches more nearly to the European species; as if, as is believed to have been the case with so many of our wild animals, the first pioneers came across from Asia by way of Behring Strait to spread southward and eastward through the Canadian woods, and southward along the mountain regions, at last down across the United States;

while another branch pushed westward along the northern border of the European forests across Siberia and northern Russia to Finland and Poland, but apparently without penetrating southward anywhere as they have in this country.

In Asia there are said to be a dozen or more distinct species, but these two, the European and American, appear to be the only ones that have strayed beyond its borders. I have sometimes been tempted to believe that the Canadian variety is occasionally found much farther south than is commonly supposed to be the case, even to southern New Hampshire, as I have obtained one or two specimens that corresponded very closely to that type both in size and colour. Half a dozen years ago or more, moreover, a local gunner told me that he had found a family of flying-squirrels in a hollow willow-tree in this town, and that he had been particularly struck by their large size and dark colour and the fact that each of them had a very noticeable broad black band along the side. I should not have put much credence in his tale if his description had not applied so closely to the northern variety, which it was hard to imagine that he could ever have heard of.

It is now several years since I have seen a live flying squirrel, though there is no reason to suppose that they are any less abundant than formerly. I have rapped on hollow trees and pried into decaying logs and stumps on every occasion without discovering the sleepy little chaps I was in search of. But this sort of thing goes largely by chance after all, and to-morrow I may happen on them where I least expect it. I remember once climbing to a crow's nest in a tall pine while the old birds wheeled and scolded overhead. When rather more than half-way to the top, I reached the place that I had seen from the ground, but was disappointed to find only a last year's nest heaped up with dry leaves and pine-needles in such a way as to show that it had already been appropriated by squirrels. On investigation, I found, instead of red squirrels as I had expected, four or five little flying squirrels about half-grown. I only saw them for a few seconds at most, as they scrambled away in all directions and disappeared completely. But in those few seconds I became aware that young flying squirrels are simply the most delightful things in existence. And I still look forward to the time when I shall discover another family of them, without the slightest fear of being disenchanted.

www.ingramcontent.com/pod-product-compliance
Lightning Source LLC
Chambersburg PA
CBHW021054030726
47496CB00006B/1837